The Spatial Humanities

SPATIAL HUMANITIES

David J. Bodenhamer, John Corrigan, and Trevor M. Harris, editors

The spatial humanities is a new interdisciplinary field resulting from the recent surge of scholarly interest in space. It prospects a ground upon which humanities scholars can collaborate with investigators engaged in scientific and quantitatively-oriented research. This spatial turn invites an initiative focused on geographic and conceptual space and is poised to exploit an assortment of technologies, especially in the area of the digital humanities. Framed by perspectives drawn from Geographic Information Science, and attentive to cutting-edge developments in data mining, the geo-semantic Web, and the visual display of cultural data, the agenda of the spatial humanities includes the pursuit of theory, methods, case studies, applied technology, broad narratives, persuasive strategies, and the bridging of research fields. The series is intended to bring the best scholarship in spatial humanities to academic and lay audiences, in both introductory and advanced forms, and in connection with Web-based electronic supplements to and extensions of book publication.

EDITORIAL ADVISORY BOARD

Edward L. Ayers, University of Richmond, USA
Peter Bol, Harvard University, USA
Peter Doorn, DANS, Netherlands
I-chun Fan, Academia Sinica, Taiwan
Michael Goodchild, University of California-Santa Barbara, USA
Yuzuru Isoda, Ritsumeikan Asia Pacific University, Japan
Kim Knott, University of Leeds, UK
Anne Knowles, Middlebury College, USA
Andreas Kunz, Institute of European History (Mainz), Germany
Lewis Lancaster, University of California-Berkeley, USA
Gary Lock, University of Oxford, UK
Barney Warf, Kansas University, USA
May Yuan, Oklahoma University, USA

The Spatial Humanities

GIS AND THE FUTURE OF
HUMANITIES SCHOLARSHIP

EDITED BY

David J. Bodenhamer
John Corrigan
Trevor M. Harris

INDIANA UNIVERSITY PRESS

Bloomington & Indianapolis

This book is a publication of

Indiana University Press
601 North Morton Street
Bloomington, IN 47404-3797 USA

www.iupress.indiana.edu

Telephone orders   800-842-6796
Fax orders   812-855-7931
Orders by e-mail   iuporder@indiana.edu

© 2010 by Indiana University Press
All rights reserved

No part of this book may be reproduced or utilized in any form or by any means, electronic or mechanical, including photocopying and recording, or by any information storage and retrieval system, without permission in writing from the publisher. The Association of American University Presses' Resolution on Permissions constitutes the only exception to this prohibition.

⊚The paper used in this publication meets the minimum requirements of the American National Standard for Information Sciences—Permanence of Paper for Printed Library Materials, ANSI Z39.48-1992. Manufactured in the United States of America

Library of Congress Cataloging-in-Publication Data

The spatial humanities : GIS and the future of humanities scholarship / edited by David J. Bodenhamer, John Corrigan, and Trevor M. Harris.
    p. cm. — (Spatial humanities)
  Includes bibliographical references and index.
  ISBN 978-0-253-35505-8 (cloth : alk. paper) — ISBN 978-0-253-22217-6 (pbk. : alk. paper) 1. Geographic information systems—Social aspects. 2. Human geography. 3. Humanities—Social aspects. 4. Humanities—Social aspects—Methodology. 5. Memory—Social aspects. 6. Learning and scholarship—Technological innovations. I. Bodenhamer, David J. II. Corrigan, John. III. Harris, Trevor.
  G70.212.S654 2010
  001.30285—dc22
                                    2009053214
1 2 3 4 5 15 14 13 12 11 10

Contents

· Introduction   vii

1 Turning toward Place, Space, and Time   1
  Edward L. Ayers

2 The Potential of Spatial Humanities   14
  David J. Bodenhamer

3 Geographic Information Science and Spatial
  Analysis for the Humanities   31
  Karen K. Kemp

4 Exploiting Time and Space: A Challenge for
  GIS in the Digital Humanities   58
  Ian Gregory

5 Qualitative GIS and Emergent Semantics   76
  John Corrigan

6 Representations of Space and Place in the Humanities   89
  Gary Lock

7 Mapping Text   109
  May Yuan

8 The Geospatial Semantic Web, Pareto
  GIS, and the Humanities   124
  Trevor M. Harris, L. Jesse Rouse, and Susan Bergeron

9 GIS, e-Science, and the Humanities Grid   143
  Paul S. Ell

10 Challenges for the Spatial Humanities:
   Toward a Research Agenda   167
   Trevor M. Harris, John Corrigan, and David J. Bodenhamer

· *Suggestions for Further Reading*   177

· *List of Contributors*   191

· *Index*   195

Introduction

This book proposes the development of spatial humanities that promises to revitalize and redefine scholarship by (re)introducing geographic concepts of space to the humanities. Humanists are fully conversant with space as concept or metaphor—gendered space, the body as space, and racialized space, among numerous other rubrics, are common frames of reference and interpretation in many disciplines—but only recently have scholars revived what had been a dormant interest in the influence of physical or geographical space on human behavior and cultural development. This renewal of interest stems in large measure from the ubiquity of Geographic Information Systems (GIS) in contemporary society. From online mapping and personal navigation devices to election night maps colored in red and blue, we are more aware than ever of the power of the map to facilitate commerce, enable knowledge discovery, or make geographic information visual and socially relevant.

GIS lies at the heart of this so-called spatial turn. At its core, GIS is a powerful software that uses location to integrate and visualize information. Within a GIS, users can discover relationships that make a complex world more immediately understandable by visually detecting spatial patterns that remain hidden in texts and tables. Maps have served this function for a long time: the classic example occurred in the 1850s when an English doctor, John Snow, mapped an outbreak of cholera and saw how cases clustered in a neighborhood with a well that, unknown to residents, was contaminated. Not only does GIS bring impressive computing power to this task, but it is capable of integrating data from different formats by virtue of their shared geography. This ability has attracted considerable interest from historians, archaeologists, linguists, students of material culture, and others who are interested in place, the dense coil of memory, artifact, and experience that exists in a particular space, as well as in the coincidence and movements of people, goods, and ideas that have occurred across time in spaces large and small. Recent years have witnessed the wide application of GIS to historical and cultural questions: did the Dust Bowl of the 1930s result from over-farming the land or was it primarily the consequence of long term environmental changes? What influence did the rapidly changing cityscape of London have on literature in Elizabethan England? What was the relationship between rulers and territory in the checkered political landscape of state formation in nineteenth-century Germany? How did spatial networks influence the administrative geography of medieval China? Increasingly, scholars are turning to GIS to provide new perspective on these and other topics that previously have been studied outside of an explicitly spatial framework.

Spatial humanities, especially with a humanities-friendly GIS at its center, can be a tool with revolutionary potential for scholarship, but as such, it faces significant obstacles at the outset. The term humanities GIS sounds like an oxymoron both to humanists and to GIS experts. It links two approaches to knowledge that, at first glance, rest on different epistemological footings. Humanities scholars speak often of conceptual and cognitive mapping, but view geographic mapping, the stock in trade of GIS, as an elementary or primitive approach to complexity at best or environmental determinism at worst. Experts in spatial technologies, conversely, have found it difficult to wrestle slippery humanities notions into software that demands precise locations and closed polygons. At times, applying GIS to the humanities appears only to prove C. P. Snow's now-classic formulation of science and the humanities as two separate worlds.

One of the problems, perhaps the basic problem, is that GIS was not developed for the humanities. It emerged first as a tool of the environmental sciences. Oriented initially around points, lines, and polygons, it found quick acceptance in the corporate world and, with its close cousin, GPS, spawned a host of location-based services. Its uptake in the academy was slower, although by the 1980s it was possible to speak of a "spatial turn," a re-emergence of space and place as important concepts in the social sciences, driven in large measure by GIS and other spatial technologies. Humanities too experienced a spatial turn—and a temporal turn in the

New Historicism—but its spaces and places were metaphorical rather than geographical constructions. Although GIS has gained a small foothold in specialty areas such as historical GIS, the technology that drove a social science agenda for two decades had little salience for humanists, who saw scant potential in it for answering the questions that interested them.

Significantly, the discipline that provided the home for much GIS development and application, geography, found itself divided over the technology in ways that mimicked the concerns expressed by humanists about quantitative methods generally. The central issue was, at heart, epistemological: GIS privileged a certain way of knowing the world, one that valued authority, definition, and certainty over complexity, ambiguity, multiplicity, and contingency, the very things that engaged humanists. From this internal debate, often termed Critical GIS, came a new approach, GIS and Society, which sought to reposition GIS as GIScience, embodying it with a theoretical framework that it previously lacked. This intellectual restructuring pushed the technology in new directions that were more suitable to the humanities. The aim of this book is to seize the momentum generated by the long debate in geography and use it to advance an even more radical conception of GIS that will reorient, and perhaps revolutionize, humanities scholarship.

The power of GIS for the humanities lies in its ability to integrate information from a common location, regardless of format, and to visualize the results in combinations of transparent layers on a map of the geography shared by the data. Internet mapping has made this concept widely recognized and accessible, but this use of GIS only hints at its potential for the humanities. Scholars now have the tools to link quantitative, qualitative, and image data and to view them simultaneously and in relationship with each other in the spaces where they occur. But the technology currently requires that humanists fit their questions, data, and methods to the rigid parameters of the software, which implicitly are based on positivist assumptions about the world. We seek instead to conceptualize spatial humanities by critically engaging the technology and directing it to the subject matter of the humanities, taking what GIS offers in the way of tools while at the same time urging new agendas upon GIS that will shape it for richer collaborative engagements with the humanistic disciplines. It will not be sufficient for the humanities to draw piecemeal from the vocabulary of spatial analysis redolent in GIS or simply to adapt the current state of GIS technology to specific research. Rather, genuine advancement of scholarly investigation of space in the humanities will derive from investigators' successes in effecting a profound blending of research languages and in organizing sustained collaborative experimentation with spatially aware interpretation.

To date, studies using GIS in historical and cultural studies have been disparate, application driven, and often tied to the somewhat more obvious use of GIS in census boundary delineation and map making. While not seeking to minimize the importance of such work, these studies have rarely addressed the broader, more fundamental issues that surround the introduction of a spatial technology such as GIS into the humanities. There are core reasons why GIS has found early use and ready acceptance in the sciences and social sciences rather than in the more qualitatively based humanities. The humanities pose far greater epistemological and ontological issues that challenge the technology in a number of ways, from the imprecision and uncertainty of data to concepts of relative space, the use of time as an organizing principle, and the mutually constitutive relationship between time and space. Essentially, GIS and its related technologies currently allow users to determine a geography of space. In the context of the humanities, we seek to move GIS from this more limited quantitative representation of space to facilitate an understanding of place within time and the role that place occupies in humanities disciplines.

Seeking to fuse GIS with the humanities is challenging in the extreme. GIS is a technology that generates geometric abstractions of the real world that can be mathematically integrated to provide a powerful spatial analytic system. Such a positivist science sits uncomfortably with the varied philosophical and methodological approaches traditionally pursued in the humanities. The qualitative-based humanities are problematic for a quantitative technology. Quantitative representations of space fit more comfortably with the sciences and social sciences than they do with qualitatively based humanities. GIS is spatially deterministic and requires landscapes and societal patterns and processes to be tied to the spatial geometrical primitives of point, line, polygon, and pixel. The mathematical topology that underpins GIS brings its own data representations in the form of raster, vector, and object forms. The attribution of these geometric forms lends itself to the classifications of natural resources, infrastructure, demography, and environmental phenomena rather than to the less well-defined descriptive terms and categories of the humanities. Spatio-temporal GIS, or the ability of GIS to handle space and time concurrently, also remains unresolved, which makes current technology difficult for time-based humanities studies. Data and the representations of phenomena, then, are singular factors that challenge the fusion of GIS with the humanities. Yet the GIS abstractions of space, nature, and society, while posing substantial problems, are particularly relevant in the humanities where notions and representations of place, rather than those of space, are primary. To this end, GIScientists have made recent advances in spatial multimedia, in GIS-enabled Web services, geovisualization, cyber geography, exploratory spatial data analysis, and virtual reality that provide capabilities far exceeding the abilities of GIS on its own. Together, these technologies have the potential to revolutionize the role of place in the humanities by moving beyond the two-dimensional map to explore dynamic representations and interactive systems that will prompt an experiential, as well as rational, knowledge base.

This notion of a richer, dynamic, and experiential GIS resonates with the evocative and thick descriptions of place and time that humanists have long favored in their scholarship. Even mapping itself comports well with the aims and methods of humanists. Representation of the past, suggests historian John Lewis Gaddis, is a kind of mapping where the past is a landscape and history is the way we fashion it. The metaphor, one consistent with disciplinary traditions across the humanities, makes the link between "pattern recognition as the primary form of human perception and the fact that all history . . . draws upon the recognition of such patterns."[1] In this sense, mapping is not cartographic but conceptual. It permits varying levels of detail, not just as a reflection of scale but also of what is known at the time. Like the map, history becomes better and more accurate as we continue to accumulate more detail, observe its patterns, and refine our knowledge.

This conception of history may be applied to the humanities more generally. Humanists, as the term implies, study the human condition in all its variety. The various disciplines in the humanities have their own traditions, of course, but collectively they would agree that their aim is to present a reasoned argument about the known past that allows us to learn who we are and what we may become as individuals, groups, and societies. This inquiry is part of our nature as humans, but it comes fraught with difficulties. The past, of course, is irretrievable, which is why historians draw a sharp distinction between it and the arguments or lessons we derive from it. We understand the past's value: it is our source of evidence; without it, we would know nothing or have any sense of who we are. But the past escapes us as soon as it becomes past. We cannot recapture it; we can only represent it. In representing the past, we seek perspective, the point of view that allows us to discern patterns among the events that have occurred. We are not trying to transmit accumulated knowledge—culture and tradition do this, among other means—but to understand the significance of our experience.

In their essence, humanities disciplines seek to generalize from the particular, not for the purpose of finding universal laws but rather to glean insights about cause and effect from a known outcome. Here, the humanities differ from much social science, which attempts to reach a generalization that holds true in any similar circumstance. This difference is significant and influences the way the two groups of scholars create knowledge. For many social scientists, the search for trustworthy generalization focuses on the isolation of an independent variable, the cause that has a predictable effect on dependent variables or ones that respond to the stimulus or presence of a catalyst. They believe it is possible to discover such a variable, given sufficient resources, because the world is not yet lost to them. Humanists must contend with fragmentary evidence and are painfully aware that the past is incomplete and irretrievable. They also are skeptical of prediction. The past is fixed, even if its interpretation is not: in it the intersection of patterns and singular events can be discovered. Not so in the future, where continuities and contingencies coexist independently of one another. Humanists view reality as web-like, to use philosopher Michael Oakeshott's phrase, because they see everything as related in some way to everything else. Interdependency is the lingua franca of the humanities.

Humanists seek to portray a world that is lost for the purpose of answering questions that bear on human experience as we perceive it today. The humanities scholar's goal is not to model or replicate the past; a model implies the working out of dependent and independent variables for purposes of prediction, whereas replication suggests the ability to know the past and its cultural forms more completely than most humanists would acknowledge is possible. Humanists, in a sense, are abstractionists: they have the capacity for selectivity, simultaneity, and the shifting of scale in pursuit of the fullest possible understanding of heritage and culture. Traditionally, humanities scholars have used narrative to construct the portrait that furthers this objective. Narrative encourages the interweaving of evidentiary threads and permits the scholar to qualify, highlight, or subdue any thread or set of them—to use emphasis, nuance, and other literary devices to achieve the complex construction of past worlds. All of these elements—interdependency, narrative, and nuance, among others—predispose the humanists to look askance at any method or tool that appears to reduce complex events to simple schemes. The computer, of course, is a technology that does not tolerate ambiguity, expressing all matter as zeroes and ones and demanding mutually exclusive categories in its data structures. Its insistence on precision does not fit the worldview of humanities scholars; indeed, these disciplines appear at times to embrace an uncertainty principle—the more precisely you measure one variable, the less precise are other variables.

It is no accident that humanists have embraced eclectic methods as fervently as they resist anything that smacks of reductionism. Questions drive humanities scholarship, not hypotheses, and the questions that matter most address causation: why matters more than whom, what, or when, even though these latter questions are neither trivial nor easy to answer. The research goal is not to eliminate explanations or to disprove the hypothesis but to open the inquiry through whatever means are available and by whatever evidence may be found. A well-presented argument often does not settle a question; it may complicate it or open new questions that previously were unimagined. Similarly, humanists are hard-pressed to identify a preferred method because each avenue of investigation yields different evidence and thus different insights. We revisit evidence as we discover new data. Our approach is recursive, not linear: our goal is not so much to eliminate answers as to admit new perspectives. The evidence we use, even if fraudulent, is rarely discarded because it may answer another question: the data may be false but it also raises the question of why it exists and what significance does it have on an understanding of human behavior. Our approach to problems doubtless appears quixotic to non-humanists because it does not lead to finality. But for humanists, the goal is not proof but meaning. The challenge, then, for humanities GIS is to use technology to probe, explore, challenge, and complicate, in sum, to allow us to see, experience, and understand human behavior in all its complexity. As in traditional humanities scholarship, the goal is less to produce an authoritative or ultimate answer than to prompt new questions, develop new perspectives, and advance new arguments or interpretations.

How we do this is one of the large aims of this book. The authors—three historians, a religionist, an archaeologist, and four geographers: three scholars from the UK, five from the U.S.—were participants in an expert workshop held in June 2008 at The Polis Center, a research unit of the IU School of Liberal Arts at Indiana University Purdue University Indianapolis (IUPUI). Held under the aegis of the Virtual Center for Spatial Humanities, a partnership among Florida State University, West Virginia University, and IUPUI, the sessions sought to develop an interdisciplinary framework and language for a spatial and visual approach to the humanities, as well as identify the potential of GIS and GIScience to contribute meaningfully to humanities scholarship. Working from papers prepared in advance and revised for this volume, the participants grappled with theories and technologies, concepts and critiques, potential and practicalities. These chapters represent different sorts of queries about what currently is possible in exploring space in the humanities and about where promising frontiers are opening, especially in terms of our ability to adapt technology to new ends. It is our hope that they provoke creative thinking about how the technology that organizes knowledge on the Web and renders space visually in GIS can be shaped in ways that better accommodate the methods and approaches of the humanities, leading to broader and deeper collaboration between humanists and geographical information scientists.

What resulted from the workshop were the chapters in this book, as well as the beginnings of an agenda for research that will test what we discussed against major questions of interest to humanists. The essays that follow grapple with problems—how to create a language that bridges disciplines, how to re-conceptualize the humanities to include spatial perspectives, how to use GIS to analyze texts and images as well as it parses points and polygons—and suggest approaches toward a robust spatial humanities. From deep mapping to immersive GIS, the solutions discussed in these pages are well within reach. They rely on the rapid convergence of technologies that marks Web 2.0, but they also are grounded in well-established theories of critical geography and postmodern humanities. It is our hope—indeed, our expectation—that these essays will prompt a long-needed reintegration of geography into the humanities. All of our disciplines will be richer for such an outcome.

The editors gratefully acknowledge the National Endowment for the Humanities, Indiana University, Florida State University, and the Eberly College of Arts and Sciences of West Virginia University for financial support in preparation of this book and the series of which it is a part.

DJB, JC, and TMH

NOTE

1. John Lewis Gaddis, *The Landscape of History: How Historians Map the Past* (New York: Oxford University Press, 2002), 33.

The Spatial Humanities

ONE

Turning toward Place, Space, and Time

EDWARD L. AYERS

Just as many disciplines rediscovered place and space over the last thirty years, so did they rediscover time and temporal representation. A critical geography and a new historicism have reoriented many humanists and social science disciplines. Like the spatial turn, the temporal turn now grounds the analysis of everything from literature to sociology in new kinds of contexts. The exciting challenge before us now is integrating those new perspectives, taking advantage of what they have to teach us.

The spatial turn began within the discipline of geography itself. By the early 1970s, geographer Edward Soja observes, many people in the field "sought alternative paths to rigorous geographical analysis that were not reducible to pure geometries." In this new critical geography, "rather than being seen only as a physical backdrop, container, or stage to human life, space is more insightfully viewed as a complex social formation, part of a dynamic process." By making this argument, geographers opened their discipline to humanists and social scientists who found congenial both a skepticism toward positivist social science and a focus on the texture of experience.[1]

For non-geographers, the spatial turn has been largely defined by a greater awareness of place, manifested in specific sites where human action takes place. As Karen Halttunen told the members of the American Studies Association in her presidential address, studies of place in the humanities have tended to focus on the particular, the narrative, and the concrete, to show "a strong sense of the constructedness of place, of place-making as an ongoing and always contested process, and of the creative variety of cultural practices employed for placemaking." In the 1970s and 1980s, Halttunen noted, "spatial analysis tended to the metaphorical, as we adopted the idiom of borders and boundaries, frontiers and crossroads, centers and margins. In literature, the new regionalism and the booming field of ecocriticism foreground what had been considered mere background or setting."[2]

At the same time the critical geography of place gained momentum, another innovation in geography sped up as well. This geography, based in the rigorous mathematical background of many practitioners in the field, grew from new technological developments, especially Geographic Information Systems (GIS). Not content simply to apply the new tools, however, "geographers became increasingly concerned with the fundamental theoretical issues related to spatial data handling," geographer Daniel Sui points out. "Geographers were no longer intellectually satisfied with mere technical innovations. If GIS had become the answer, many geographers were itching to ask, what was the question?"[3]

Geographers have used cognitive science, computer science, physics, non-Euclidean geometry, neural computing, and fractal geometry to extend their understanding of space, with each new method and conjunction of methods raising new questions. Greater power of analysis brings new questions to the surface. GIS, Sui observes, "has been examined through every critical lens of social theory and poststructuralist perspective, ranging from feminist theories to indigenous knowledge, public participation, hermeneutics, political ecology, actor-network theory, critical media theories, and linguistic philosophies and ethics." Sui believes that "computational, spatial, social, environmental, and aesthetic dimensions" will all flourish as geography moves forward, for geography "is a fertile ground for crossing the traditional boundaries of science, social theory, technology, and the humanities."[4]

The study of place and the study of space, in other words, converge in a heightened self-awareness that is useful for geographers and others as well. "Because of geography's focus on studying subject matter in common with the humanities and sciences or the human and natural sciences, it has sometimes been called the bridging discipline or an interfacing or fusing discipline," geographer Stanley Brunn argues. "That is, it is the discipline most concerned with studying the relationships between the human and physical phenomena." Geographers "are both exporters and importers of knowledge" and thus geography serves as a sturdy bridge crossed by many disciplines.[5]

Another bridging discipline deals with the other defining context in human life: time. That discipline, of course, is history. Maps and history are deeply complementary. "Both reduce the infinitely complex to a finite, manageable, frame of reference," theorist Denis Cosgrove points out. "Both involve the imposition of artificial grids—hours and days, longitude and latitude—on temporal and spatial landscapes, or perhaps I should say timescapes and landscape. Both provide a way of reversing divisibility, of retrieving unity, of recapturing a sense of the whole, even though it can not be the whole." Maps and histories do the same kind of work in different disciplines, in different dimensions of human experience.[6]

History, no less than other disciplines, took its own spatial turn. The turn did not prove a wrenching change of direction for history, because history has always had a strong spatial component. Historians have long relied on maps and have always plotted stories in space as well as time. Geographers and historians have usually seen the other as allies, fellow travelers. That is because, as D. W. Meinig, a pioneering practitioner of both disciplines, argues, "geography, like history and unlike the sciences, is not the study of any particular kind of thing, but a particular way of studying almost anything. Geography is a point of view, a way of looking at things. If one focuses on how all kinds of things exist together spatially, in areas, with a special emphasis on context and coherence, one is working as a geographer." And if we substitute "temporally" for "spatially" in the preceding sentence, and exchange "historian" for "geographer," we are describing history.[7]

History has absorbed place, the more humanistic aspect of the spatial turn, in studies of everything from regions and the environment to consumer culture and slavery. But it has not quite known what to do with the more analytical, technologically enabled component of the new geography. That is in part because history is, at heart, a humanistic discipline rather than a social science. Despite forays into quantification, history tends toward the singular and particular, toward interpretation rather than generalization, toward the narrative rather than the model. Historians, accordingly, have not developed very explicit theories of space or place. Each representation tends to be handmade, custom-built.

Despite the affinities of history and geography, trying to comprehend space, place, and time in concert has always proven difficult. As the historian Hugh Trevor-Roper asked decades ago, "How can one both move and carry along with one the fermenting depths which are also, at every point, influenced by the pressure of events around them? And how can one possibly do this so that the result is readable? That is the problem."[8] How, in other words, might we combine the obvious strengths of geographic understanding with the focus on the ineffable, the irreducible, the singular, that is at the heart of history? How might we integrate structure, process, and event? How might we combine space, place, and time?

Novelists have figured out ways to represent the concatenation of time and space in human lives. More than seventy years ago the brilliant Russian literary critic Mikhail Bakhtin defined the "chronotope," "the intrinsic connectedness of temporal and spatial relationships that are artistically expressed in literature." Bakhtin described the magic of the chronotope in beautiful language, even in translation: "In the literary artistic chronotope, spatial and temporal indicators are fused into one carefully thought-out, concrete whole. Time, as it were, thickens, takes on flesh, becomes artistically visible; likewise, space becomes charged and responsive to the movements of time, plot and history." For Bakhtin, "the chronotope is the place where the knots of narrative are tied and untied. ... Time becomes, in effect, palpable and visible; the chronotope makes narrative events concrete, makes them take on flesh, causes blood to flow in their veins."[9]

Sometimes historians can create the magical narrative effect Bakhtin describes, the same evocative fusion of place and time in human experience. But they cannot count on that success and, besides, historians have responsibilities beyond narrative. They need not only to evoke time and space, but to explain in more explicit ways the workings of both and the relationship between the two. Since time and space are so closely linked, it may be that the spatial turn can present an opportunity to think about time in new ways as well.

One line of thought, decades now in the making, the product of sociologists, anthropologists, historians, and others, seems to hold out the promise of unifying action in place and time. This tradition of analysis, written in a common spirit but not constituting a unified school, argues that social power can be best perceived in "practice" rather than in categories. These analysts of what has come to be called "practice theory" have taken things apart—dismantling generalizations about cultures, classes, races, and societies, casting aside older Marxian, neoclassical, and structuralist models—and put them back together in more dynamic, interrelated, and complicated ways. They show that the cultural and the material are parts of the same processes and structures, that they cannot be separated. Leading theorists in this vein include Pierre Bourdieu, Anthony Giddens, Marshall Sahlins, Raymond Williams, and Sherry Ortner.

Historians have not been leaders in defining practice theory, but they have recently taken up discussion of the approach. Gabrielle M. Spiegel, synthesizing the literature, argues that practice theory's accent "on the historically generated and always contingent nature of structures of culture returns historiography to its age-old concern with processes, agents, change, and transformation, while demanding the kind of empirically grounded research into the particularities of social and cultural conditions with which historians are by training and tradition most comfortable." William H. Sewell, Jr., another prominent historian, believes that "social life may be conceptualized as being composed of countless happenings or encounters in which persons and groups of persons engage in social action. Their actions are constrained and enabled by the constitutive structures of their societies." As a result, "'societies' or 'social formations' or 'social systems' are continually shaped and reshaped by the creativity and stubbornness of their human creators."[10]

This model bears a striking resemblance to notions of place portrayed after the spatial turn. Like the highly inflected, multifarious representations of space growing out of critical geography, time in practice theory is less a unified field, a background, than an active participant in the story. The sociologist Andrew Abbott describes time as others describe space. Time, for him, is "a series of overlapping presents of various sizes, each organized around a particular location and overlapping across the whole social process." Time is not fixed, not a given. "Within this complex world, change is the normal state of affairs. We do not see a largely stable world that changes occasionally, but a continuously changing world that has macroscopic stabilities emerging throughout it. This world is a world of events." Stability in this eventful world is not the default. As a result, "the fact that everything—no matter how large—is perpetually being reproduced means that everything—no matter how large—is always on the line. So sudden large-scale change is not surprising."[11]

Time, like geography, can be disassembled analytically. Just as we differentiate between a more generalized space and a more localized place, so can we differentiate general processes from specific events. We live daily in places and events but we are parts of large spaces and processes we can perceive through efforts of disciplined inquiry. Just as a geographer relates place and space, so do historians relate event and process. Geography locates us on a physical and cultural landscape while history locates us in time. Joining the two kinds of analysis in a dynamic and subtle way offers an exciting prospect. Practice theory, a supple way to imagine both structure and activity, may help.

The everyday and the local, a common focus in practice theory, would seem to have at least one great limitation: explaining larger social changes. How do we get from the prosaic to the transformational? In fact, practice theory proves to be a way to explain how big and sudden changes penetrate deeply into people's hearts and minds. "All social life is 'contingent,' implicated and unpredictable, because all parts of life depend on each other," I have argued elsewhere. "What we think of as public and private, economic and political, religious and secular, and military and civilian are deeply connected. Social change can start anywhere and lead anywhere." Such a perspective argues for the intricate interplay of the structural and the ephemeral, the enduring and the emergent. This is "deep contingency," a view of social life that fuses an active sense of place and an active sense of time.[12]

Deep contingency tries to suggest how societies can change their self-understanding quickly and profoundly. Secession in the United States, where states decided in a matter of weeks to join a new Confederacy and sacrifice everything in that new purpose, is one example; others might include the Russian Revolution or the fall of the Berlin Wall. Practice theory addresses these ruptures. As William Sewell argues, "big and ponderous social processes are never entirely immune from being transformed by small alterations in volatile local social processes. . . . Because structures are articulated to other structures, initially localized ruptures always have the potential of bringing about a cascading series of further ruptures that will result in structural transformations—that is, changes in cultural schemas, shifts of resources, and the emergence of new modes of power."[13]

Deep contingency needs to be distinguished from what we might call surface contingency, the familiar historical staples of accident, personality, and timing, the clichés of "what ifs" and "almosts." While surface contingency can sometimes trigger deep contingency, the great majority of unpredictable events come and go without much consequence; deep contingency, visible only after it has arrived, reverberates throughout the recesses of the social order. "A single, isolated rupture rarely has the effect of transforming structures because standard procedures and sanctions can usually repair the torn fabric of social practice," Sewell argues. "Ruptures spiral into transformative historical events when a sequence of interrelated ruptures disarticulates the previous structural network, makes repair difficult, and makes a novel rearticulation possible."[14]

To understand deep contingency we must try to comprehend a society as a whole, its structures of ideology, culture, and faith as well as its structures of economics and politics. All structures must be put into motion and motion put into structures. As literary scholar Raymond Williams insists, "Determination of this whole kind—a complex and interrelated process of limits and pressures—is in the whole social process itself and nowhere else: not in an abstracted 'mode of production' nor in an abstracted 'psychology.'" Or, as anthropologist Sherry Ortner explains, "A practice approach has no need to break the system into artificial chunks like base and superstructure (and to argue over which determines which), since the analytic effort is not to explain one chunk of the system by referring to another chunk, but rather to explain the system as an integral whole (which is not to say a harmoniously integrated one) by referring it to practice." And, of course, space and time are crucial components of that integral whole.[15]

By its very nature deep contingency depends on larger processes, on interconnected systems. Portrayals of particular places, often apprehended through the finely grained portrayals of a case study, struggle to convey what we might be able to see on a broader canvas. Deep contingency cascades throughout a society, but it has to start somewhere, often in political or economic decisions made in capitals or metropoles. Mapping offers a way to see deep contingency in motion, rippling and sweeping across space and time.

New thinking in geography, history, and theory, combined with new technology and techniques, suggests that we may be able to represent the intersection of space and place, process and event in more compelling ways. Some of this representation, as in fiction, may take place in writing, in new kinds of narratives sensitive to the ways time and place interact and intersect. Other representation may be possible with new technologies that permit us to integrate various aspects of human experience in more flexible ways.

My colleagues and I at the University of Richmond are working on two related but separate experiments to see how it might be possible to relate geography and history more clearly. Both experiments take advantage of relatively straightforward and inexpensive computing techniques deployed in unusual ways. Both of the experiments marry the techniques of humanists with those of other disciplines, focusing less on causation (claims about which humanists are rightly wary) and more on interpreting the consequences and resonances of events. By laying down grids of space and time and documenting the actions of people across those grids, we are able to see patterns we could not see otherwise.

The first experiment is called the History Engine. The goal of this tool is to capture and convey the richness, the particularity and singularity, of both place and time while allowing us to see larger patterns otherwise invisible. The History Engine is a moderated wiki, populated thus far by hundreds of students at five colleges and universities. The project's Web site describes the technique: "Student participants research primary documents and use secondary sources to help reconstruct the 'episodes'—snippets of daily life from the largest national event to the smallest local occurrence—that make up the cumulative database." Students do the work of historians. They "examine primary documents, place them in a larger historical context using secondary resources, and prepare their analysis for the public eye." The elements in the searchable database undergo "a careful academic screening process on the part of library staff, archivists, professors, and teaching assistants. Because only registered students can contribute, each episode is carefully vetted for content and accuracy." Once approved, the episodes are added to others until the database is populated by thousands of documented nodes where place, time, and action intersect, searchable in several dimensions.[16]

The episodes in the History Engine embody the principles of practice theory. The episodes demonstrate how people enact the dramas of their society in places large and small, confronting common challenges and opportunities, each episode unique and yet part of larger patterns. The History Engine shows that, at base, history is where singular events and larger patterns intersect. One can watch the secession crisis of 1861, for example, move like a wave through the words and experiences of people scattered across an area the size of continental Europe, the momentum building on itself.

Each episode in the History Engine, moreover, is geocoded so that it can be represented in space as well as in time. A user may start with a map of the United States, showing the density of episodes in each county in each decade, as well as with a search of keywords or dates. Each episode is embedded, in other words, in place and time. The History Engine shows how pattern, structure, event, and change are embodied at the local and personal level, in a collage of moments. It is, as it were, analog, requiring interpretive acts of translation. The History Engine can convey primary documents, images, and multimedia elements as well as episodes and so could serve as a vehicle for deeply layered and textured kinds of interpretive history.

Another way of approaching place and space, event and process, involves pulling the camera back to see larger patterns in motion. This strategy, useful for looking across broader arrays of space and time, draws more on the machine-aided capacities of GIS and other digital tools.

The first model we have built at Richmond focuses on voting. It "examines the evolution of presidential politics in the United States across the span of American history. The project offers a wide spectrum of cinematic visualizations of how Americans voted in the presidential elections at the county level." It presents maps that morph from one election to the next in a fluid movement, representing a series of perspectives: raw voting distributions, political party strength, third-party challenges, highly contested counties, voter turnout rates, demographic changes, and population densities. Built with GIS but using special effects software to create the impression of motion—not unlike weather maps that show fronts, storms, and clear areas—the maps offer a new perspective on American politics.[17]

The cinematic maps are something like time-lapse photography of plants opening, of leaves unfurling in particular shapes, of vines reaching to grasp nearby structures, of diseased or thwarted processes. Or perhaps they could be imagined as models of streams and rivers, with currents folding back on themselves, of flows around submerged objects.

The emergent patterns can not be easily perceived in static maps. It is possible, in fact, that people simply do not have the neural bandwidth to deal with space and time simultaneously, in the same cognitive space, without the tricks of narrative or the aid of machinery. People tend to think of cause and effect in linear forms because that is how we get through the daily acts of life. We seem able only to tell ourselves one story at a time, to unfold one sequence in our heads at a time. We cannot picture simultaneity or envision complex processes without at least writing things down or, better, drawing pictures, or better yet, creating moving pictures.

My colleague Cindy Bukach, a cognitive neuroscientist, points out that "our perceptual system is not designed to perceive the passage of time, but it is designed to see the movement of objects through space. By converting time to motion, we can visualize the passage of time (as one does as one watches the hands of a clock move). This same principle can operate not only on the scale of seconds, minutes and hours, but also on the scale of years."

Our brains like seeing these patterns, it seems, because maps of time take advantage of our "multimodal cognitive system." Motion and temporal sequencing are key to our constant triangulation of causation. As Bukach points out, "these dynamic patterns can be simultaneous, allowing inferences of common causes, or they can be sequential, suggesting causal relationships. Motion captures attention. Displaying historical information in a motion map guides the viewers' attention to changes in a somewhat automatic way, guiding even the most naïve observer to perceive the relevance of emerging trends and relationships."[18]

A famous experiment showed how crucial time and motion are to human cognition. A researcher "placed lights at the major points of people dressed completely in black and photographed in the dark. If these people did not move, observers reported seeing no identifiable pattern," Donna Peuquet reported. "If, however, the people moved in performing some ordinary activity, such as walking or dancing, the observers immediately were able to identify the appropriate number of people engaged specifically in that activity. If the people then stop moving, the observers reported that the lights returned to a random pattern, with the people seemingly disappearing." The researchers "concluded that the perception of a gestalt pattern of an event progressing in time is basic to human cognition." Our maps try to take advantage of this capacity and desire of the human brain.[19]

Elections are virtually stage sets for dramatizing the force of event and personality on subsequent occurrences and structures. Votes are clear markers of people's beliefs and actions at specific points in time and space, conducted on a regular and frequent basis. On the cinematic maps, accordingly, one can see the consequences and patterns of secession, Reconstruction, Populism, Progressivism, socialism, disfranchisement, the New Deal, the Dixiecrats, the Voting Rights Act, George Wallace, the Republicans' Southern Strategy, or Ralph Nader.

The patterns, intricate and shifting, are too complex to explain easily in words or even numbers. We can see more in the maps than we can easily say. As Peuquet points out, "the linear (i.e., one dimensional) nature of language is ill-suited to represent the higher dimensionality of a spatial information." Because "the number of terms in natural language for expressing topological spatial and temporal relationships is hard to add to and very limited," we are limited in what we can describe. As Peuquet wryly challenges, "try verbally describing the shape of Canada or the United States."[20]

The maps' complex patterns belie easy generalization in numbers as well as in words. The convenient division between red states and blue states, it becomes obvious, is profoundly misleading. That convention has implications that reach all the way down from the state to the county to the individual voter, with a powerful set of assumptions built in: people vote in ways that tend to be generally static, unreflective, and bundled. Thus, the common nomenclature of red and blue assumes that people who live near one another tend to share political ideas because of their common history, ethnicity, and economic experience. Those identities, the red/blue dichotomy assumes, tend to change slowly.[21]

Dynamic maps show, by contrast, how shifting and complex the patterns of party voting can be. Dissolving the large blocks of the electoral college maps into more precise points reveals that states are often deeply divided. Running the cursor over each county in the United States in the interactive maps shows the extent to which individual counties embodied these divisions. Some maps show the power of state boundaries, as places with the same ethnic patterns and economies vote in quite different ways because of the power of patronage, a popular governor, or scandals. Other maps may show patterns that cross state boundaries, as when counties of Appalachia, north and south, voted against Barack Obama in 2008. The concept of red and blue states, while mattering for the electoral college and therefore of great instrumental meaning, turn out to be poor indicators of the complexities of American voting.

Combined, the movement, manipulability, and specificity of the dynamic maps give us a glimpse of what deep contingency might look like over time. By allowing us to see space and time at a distance, in relatively abstract ways, the maps show us dissolving and crystallizing patterns otherwise invisible in rows of numbers or static maps based on the same data. Running the maps backward and forward show one area of activity after another, revealing new detail in each viewing. Explanatory text and video, contexts moving along with the maps, provide viewers with dynamic frameworks of narrative understanding.

Now that so many disciplines have taken both the spatial turn and the temporal turn, it will be interesting to see where and how far the branching roads may take us. Translating complex patterns into language—essential for the humanities—will be an on-going challenge. Maps or timelines, dynamic or otherwise, do not speak for themselves. Inventing new forms of interpretation and explication will be a thrilling but difficult task. Fortunately, recent history suggests we will rise to the occasion. Geographers, historians, and their allies have come a long way in a short time, crossing many bridges and borders once thought closed.

NOTES

1. Edward Soja, "In Different Spaces: Interpreting the Spatial Organization of Societies," *Proceedings* 3rd International Space Syntax Symposium (Atlanta 2001).

2. Karen Halttunen, "Groundwork: American Studies in Place"—presidential address to the American Studies Association, November 4, 2005, *American Quarterly* (March 2006) 58: 1–15.

3. Daniel Sui, "GIS, Cartography, and the 'Third Culture': Geographic Imaginations in the Computer Age," *The Professional Geographer* (2004) 56:1, 62–72.

4. Ibid. For a helpful reminder that "neither space nor place is simply something that happens out in the world, but rather that both are methods that social analysts apply is setting out to study the world," and that "we do not suppose that place is restricted to small-scale, face-to-face interaction, or space to social networks and macro-level circuits," see Richard Biernacki and Jennifer Jordon, "The Place of Space in the Study of the Social," in P. Joyce, ed., *The Social in Question: New Bearings in History and the Social Sciences* (London and New York: Routledge, 2002), 133–50. Cited here at 144.

5. Stanley Brunn, "The New Worlds of Electronic Geography," *GeoTrópico* (online), (2003) 1 (1): 11–29.

6. Denis Cosgrove, *Mappings* (London, 1999), 32.

7. D. W. Meinig, *A Life of Learning*, Charles Homer Haskin lecture, American Council of Learned Societies occasional paper No. 19 (Philadelphia, 1992), 18.

8. Quoted in Keith Thomas, "A Highly Paradoxical Historian," *New York Review of Books*, April 12, 2007, 56.

9. Mikhail Bakhtin, *The Dialogic Imagination: Four Essays,* ed. Michael Holquist (Austin: University of Texas Press, 1981), 250.

10. Gabrielle M. Spiegel, ed., *Practicing History: New Directions in Historical Riting after the Linguistic Turn* (New York: Routledge, 2005), 23; William H. Sewell, Jr., *Logics of History: Social Theory and Social Transformation* (Chicago: University of Chicago Press, 2005), 100–102, 110–111.

11. Andrew Abbott, *Time Matters: On Theory and Method* (Chicago: University of Chicago Press, 2001), 296–98.

12. Edward L. Ayers, *What Caused the Civil War? Reflections on the South and Southern History* (New York: W.W. Norton, 2005), 134–35.

13. Sewell, *Logics of History,* 227–28.

14. Ibid., 100–102.

15. Raymond Williams, *Marxism and Literature* (Oxford: Oxford University Press, 1977), 87–88; Sherry Ortner, "Theory in Anthropology since the Sixties," *Comparative Studies in Society and History* (1984) 26 (1): 148–49.

16. See http://historyengine.richmond.edu/; for the mapping component, which we will be incorporating into the next iteration, see the first version of the project at http://www.vcdh.virginia.edu/SHD/.

17. The current version of the site is at http://americanpast.richmond.edu/voting/.

18. Personal communication from Bukach to Ayers, November 15, 2007.

19. Donna Peuquet, *Representations of Space and Time,* (New York: Guilford Press, 2002), 88.

20. Ibid., 178–80.

21. See Edward Glaeser and Bryce Ward, "Myths and Realities of American Political Geography," January 2006, discussion paper number 2100, http://www.economics.harvard.edu/pub/hier/2006HIER/2100.pdf (accessed October 20, 2009).

TWO

The Potential of Spatial Humanities

DAVID J. BODENHAMER

Space is everywhere, and its definitions are legion. We are inherently spatial beings: we live in a physical world and routinely use spatial concepts of distance and direction to navigate our way through it. But this routine and subconscious sense of space is not the one that engages us as humanists. We are drawn to issues of meaning, and space offers a way to understand fundamentally how we order our world. Here, contemporary notions of space are myriad: what once was a reference primarily to geographical space, with its longstanding categories of landscape and place, is now modified by class, capital, gender, and race, among other concepts, as an intellectual framework for understanding power and society in times near and distant. We recognize our representations of space as value-laden guides to the world as we perceive it, and we understand how they exist in constant tension with other representations from different places, at different times, and even at the same time. We acknowledge how past, present, and future conceptions of the world compete simultaneously within real and imagined spaces. We see space as the platform for multiplicity, a realm where all perspectives are particular and dependent upon experiences unique to an individual, a community, or a period of time.[1] This complex and culturally relativistic view of space, the product of the last several decades, has reinvigorated geography as a discipline, just as it has engaged scholars within the humanities.

We perhaps are most aware of these intellectual currents when we contrast them to once dominant—and still popular—notions of space in the American experience. In this accounting, space in the form of land shaped the national character.[2] Compared to Europe, the American continent was vast and served as a canvas for utopian dreams and dystopian nightmares. The propagandists for settlement touted its supposed riches, describing a land ripe for conquest. Puritans and other religious settlers, who otherwise embraced it as a New Eden, also saw its wildness as another reason for an obedient and vigilant community. By the early national period, any gloomy insistence on the New World as unredeemed wilderness dimmed in comparison to republican celebrations of an American empire for liberty, with its vast open spaces the precise remedy for the crowded, freedom-denying cities of Europe. Period maps and literature alike symbolized how the great swath of accessible land was the foundation for the economic independence, democracy, and nationalism that made the nation, as Abraham Lincoln claimed, "the last, best hope of mankind." The mythology that justified westward expansion found expression in American historiography when Frederick Jackson Turner advanced his frontier thesis, an interpretation that gained currency in part because it was so unremarkable. Even in the counter-narrative that cast space as a progenitor of violence, deviance, or insularity, such as in the fiction of William Faulkner or Cormac McCarthy, conceptions of natural geography played a central role in how American imagined themselves from the earliest settlements to the last decades of the twentieth century. Space also was central to another narrative based on time, in which the new nation advanced progressively toward perfection. In this mythology, America was immune to inevitable cycles of decay or decline. Space in the form of nature was the fountain of renewal that made continued progress both possible and inevitable.

No longer does this exceptional sense of space and time dominate our national conversation, in part because we are more aware as a society of how diverse the world is but also because it has been challenged so successfully within the academy. The humanities and social sciences especially have advanced new lines of inquiry characterized by a different and more nuanced understanding of space, or, as David N. Livingstone has written, in "recent years there has been a remarkable 'spatial turn' among students of society and culture."[3] This spatial turn began in the pioneering works of social scientists such as Clifford Geertz, Erving Goffman, and Anthony Giddens and has been advanced in the humanities through the work of Michel Foucault, Michel de Certeau, Edward Said, and others whose investigation of space took the form of a focus on the "local" and on context. Subject matter once organized largely by periods increasingly embraces themes of region, disapora, colonial territory, and contact zones and rubrics such as "border" and "boundary." The shift has been accompanied by and reinforced through an equivalent concern with material culture and built environment, in observations of local representation in dress, architecture, eating, music, and other cultural markers of space and place. Climate, topology, and hydrology—all of which were important to early twentieth-century *annalistes*—likewise have reemerged as important considerations in the investigation of literatures, histories, and social and political life. As a result, our national story has become more complex and problematic. Like time, space no longer has providential meaning, but in the process it has assumed a more interesting and active role in how we understand history and culture.

Today, humanists are acutely aware of the social and political construction of space. Spaces are not simply the setting for historical action but are a significant product and determinant of change. They are not passive settings but the medium for the development of culture. All spaces contain embedded stories based on what has happened there. These stories are both individual and collective, and each of them link geography (space) and history (time). More important, they all reflect the values and cultural codes present in the various political and social arrangements that provide structure to society. In this sense, then, the meaning of space, especially as place or landscape, is always being constructed through the various contests that occur over power.[4] Consider two examples: as women have gained economic and political status, feminist geographers have called attention to how we have used gendered tropes—Mother Nature or Virgin Land—to identify space, a characterization that suggests nurture but also invites exploitation. American Indians protest, rightfully, that the vast open spaces of the New World were not wilderness but their home. There is nothing new in this development—power arrangements can be seen even in the earliest maps—but humanities scholarship increasingly reflects what may in fact by the greatest legacy of postmodernism, the acknowledgement that our understanding of the world itself is socially constructed.

Now, we face a different challenge from an unexpected quarter. An attractive and increasingly ubiquitous technology, Geographic Information Systems (GIS), suggests that the world indeed is flat, at least metaphorically, by offering a view of the physical environment seemingly stripped of its cultural assumptions. Its unparalleled ability to manage and visualize data within a spatial context has led to a rediscovery of the power of the map, although often in a peculiar, culturally uninformed way. As with many technologies, GIS promises to re-invigorate our description of the world through its manipulation and visualization of vast quantities of data by means previously beyond the reach of most scholars. Increasingly, humanists are acting on this claim, but in doing so, we again run the risk of portraying the world uncritically, this time with a veneer of legitimacy that is more difficult to detect or penetrate. GIS is a seductive technology, a magic box capable of wondrous feats, and the images it constructs so effortlessly appeal to us in ways more subtle and more powerful than words can. In our eager embrace of GIS, we have been swayed by its power but have little knowledge of how it developed or why. Yet it is this history that makes us aware of both the limits and potential of GIS for the humanities—and how much it still must change to suit our needs.

GIS emerged in the early 1960s as mapping-cum-analysis software. It emerged independently from both the Harvard Laboratory for Computer Graphics, which aimed to produce automated cartography, and the Canadian GIS, which developed computerized methods to map the land capability of Canada.[5] Its intellectual and methodological lineage is much longer than this recent past—for example, the logical overlay technique, a key feature of GIS, existed as early as the eleventh century—but what was new were powerful computers and an emergent demand from such widely distributed fields as environmental science, landscape architecture, and urban planning that prized its ability to overlay data on a map of the earth's surface. With the creation of ArcInfo®, the leading commercial package, in the 1980s, GIS quickly moved into the mainstream of computing applications and spawned a wide array of location-based services.[6]

Its movement into other parts of the academy was considerably slower. For many humanists, GIS was simply another software package, with little application to the cultural and social problems that attracted their attention. Geographers, perhaps surprisingly, found themselves divided over its value. It became the focus of quantitative geographers who saw its potential to solve spatial problems by its capacity to manage large data sets and visualize the results of spatial analysis. This latter characteristic was especially important: making data visual spurred intuitive interpretation—recognition of patterns, for instance—that remained hidden in statistical analyses. Opponents, especially human geographers, were unconvinced. As late as 1988 the president of the American Association of Geographers felt comfortable labeling GIS as "a mere technique." Tension existed between scholars who viewed the technology as the herald of a shift in scientific methodology and those who saw it as a vehicle for extending existing geographic concepts. The divide ran along a fault line increasingly known as Geographic Information Science (GIScience), a critique that GIS, although well equipped to manage quantitative spatial data, rested on a positivist and naïve empiricism and was incapable of knowledge production. Representing this view was *Ground Truth* (1995),[7] a collection of essays edited by John Pickles, a prominent critic of GIS. Collectively, the essayists expressed several concerns: technological design inevitably privileges certain conceptualizations of the world; GIS was a corporate product, designed to solve corporate problems, such as route logistics or market analysis; GIS employs a limited linear logic that is not adequate for understanding societal complexity, and as a consequence, it represents and perpetuates a particular view of political, economic, and social power.[8]

At its heart, the debate within geography rested on epistemological and ontological differences that have implications for the construction of a humanities-based GIS and GIScience. Epistemology is the branch of philosophy concerned with the theory of knowledge, or its nature and scope. Its central question—"What is knowledge?"—relates both to the essence of knowledge and to how it is produced. The latter is a methodological problem—"What is the perspective we will use to interpret entities and phenomena?"—and it is this meaning of epistemology that claims the attention of GIScientists. Ontology, the foundation of metaphysics, asks "What is real or what exists?" It studies being or existence and its basic categories and relationships. Epistemology and ontology are closely related and together they have powerful implications for our conceptions of reality. Ontology helps us classify spatial objects and relate them to each other, while epistemology provides the methodological lens we use to study the objects and their relationships. The entities we study, whether natural or man-made, exist independently of our classification, but how we identify them influences our view of reality. A pile of dirt and rock may be a mound or a mountain, and the name we give it suggests the obstacle it may pose to our movement. More significant are social classifications: for instance, poverty is a relative condition; the category of being poor depends on where we draw the line between poor and not poor. Here is where epistemology becomes important. The method we choose to interpret poverty bears heavily on how we understand its essence, what it really is.[9]

Critics argued that GIS rested on a positivist epistemology. It assumed an objective reality that we can discover through scientific method, which in this theory is the path to true knowledge. Positivism stems from the work of Auguste Comte, the nineteenth-century French philosopher widely regarded as the father of sociology. Comte suggested that the scientific method was the key to progress. Through observation and testing we are able to understand how the world operates. We then can use this true or verifiable knowledge to make predictions about the world and thereby improve it: from science comes prediction; from prediction comes action. Several problems exist with this approach, proponents of what became known as Critical GIS argued. First, the world could not be measured so precisely as positivism assumed. Knowledge was always contingent upon the perspective of the observer. Even calculations of the material world depended upon cultural assumptions; not every society accepted or used the precepts of Euclidian geometry. But GIS privileged quantitative data, which it required to be precise. It did not accept uncertainty or fuzziness. It also favored official representations of the world, a result that was highly problematic because this view reflected the influence of money and power. For purposes of economic development, for instance, local government could draw neighborhood boundaries that bore little resemblance to the community identified by residents. Finally, its use of geometric space and Boolean logic ruled out the possibility of alternate, non-Western views of the world.[10]

In practice, critics claimed, evidence about the world depends upon the perspective of the observer, a distinction that GIS obscures. Two people who view the same object may interpret it quite differently because of their different assumptions and experiences. Consider a simple example: the same body of water flowing in a channel may be called a brook, stream, or creek, depending on the region where the observer grew up. Defenders of GIS responded that this difference does not matter because, regardless of name, the object remains the same. This position epistemologically is realism. It assumes that objects exist independently of the observer: the nouns "creek," "stream," and "brook" may tell us something about the observer but they still refer to the same thing—and we can use formal rules to parse when different words refer to the same object. Supporters of the technology also rejected the charges that they were naïve in their use of GIS, arguing in turn that the software was continually evolving in an effort to solve these problems.

The early part of the twenty-first century witnessed a slackening of the debate within geography as the two camps joined under the banner of GIS and Society, forming an effort to confront the issues raised by Critical GIS. This rapprochement has led to a common acknowledgement of problems in the way GIS represents the world. GIS delineates space as a set of Cartesian coordinates with attributes attached to the identified location, a cartographic concept, rather than as relational space that maps interdependencies, a social concept. It also favors institutional or official databases as the primary source of information about the world. Both tendencies exclude non-Western conceptions of the world. Some American Indians, for example, defined the world as a set of interlinked phenomena, only some of which can be defined as geographic space.[11] It is easier to understand ancient Chinese dynasties when we see their definition of space as networks of places and actors rather than as prescribed jurisdictions with formal boundaries.[12] GIS currently has difficulty managing these different meanings of space. It remains, at heart, a tool for quantitative data, the type of evidence that admits at some level to a degree of measurement that can be replicated and verified. The precision that is necessary for statistical work does not admit readily the sort of evidence used by most humanists, and when it does, the result, usually in the form of maps, can be highly misleading, implying a certainty that the underlying evidence does not permit.

While geography grappled with the theoretical and social implications of GIS, humanists were (re)discovering space, yet the two groups took divergent paths with only occasional intersections. Although the Annales school, most notably Ferdinand Braudel, its chief practitioner, had urged scholars since the 1930s to pay attention to *géohistoire,* the linkage of geography and history, most humanists paid much less attention to the environmental context for human behavior and much more to the actions, associations, and attitudes that made a space particular, in short, a place. These places could even exist in imagined space or in memory.

The spaces of interest to the humanities also could be personal—emotional space or the body in space—and even metaphorical or fictional, a woman's place, for example, as in Virginia Woolf's story, "A Room of Her Own." Except for the *annalistes,* these spaces bore little relationship to GIS, with its emphasis on physical or geographical space. Only in two areas of the humanities—archaeology and history—did scholars begin to apply the new spatial technology and, in the process, discover its limits for their work.

Archaeologists came early to GIS, as well as to other spatial instruments such as Global Positioning Systems (GPS), in large measure because it provided a handy and more accurate toolkit for managing their research in familiar but speedier ways. Maps of uncovered human habitats, long a staple of the archaeologist, were easier to chart with the survey-based techniques of GIS. Artifacts bore a spatial relationship that was important in interpreting the past, but it was the ability to visualize past places, often in 3-D, that provided a new way to recreate past landscapes and cityscapes. Architects joined with archaeologists to create virtual worlds of ancient Rome, Jamestown in 1607, or medieval Welsh villages, for example, to test our understanding of form and function. Here, it was the ability of GIS to visualize a spatially accurate physical and man-made environment that proved the attraction. Seeing a lost landscape, reconstructing historical viewsheds, and traversing a highly detailed built environment provided insights and an experiential understanding previously unavailable as scholarship.

Historians also began to drift toward GIS, but without the intense visualization employed by archaeologists. Several early efforts centered on the development of what came to be known as spatial infrastructure, that is, the development of large quantitative data sets, such as censuses, for use within a GIS. National historical GIS projects emerged in Great Britain, Germany, the United States, China, and Russia, among others. None of these projects were inclusive of all historical periods, and many of them focused more on creating framework data for other scholars than addressing research problems. Other scholars, especially environmental historians, employed GIS to test standard interpretations by constructing a data landscape to tell a more complicated story than traditional methods allowed. Geoff Cunfer, for instance, used GIS to rebut the standard Dust Bowl narrative that blamed farmers in Oklahoma and Kansas in the 1920s and '30s for using ruinous, ecologically insensitive agricultural practices, thus turning a pristine prairie into wasteland. By mapping dust storms across a wider period and a broader scale, he concluded that, in fact, they were part of a longer-term weather and environmental pattern rather than the result of short-term human errors.[13] In a more ambitious example, Michael McCormick re-mapped Europe from AD 300 to 900, showing the connection between developments in communication and transportation that scholars previously had studied in isolation.[14] Other historians took advantage of GIS to relate data of different formats based on their common location, at times using the Internet to bring spatial and archival evidence together and allow readers to explore the evidence afresh (e.g., Valley of the Shadow Project[15] or the Salem Witch Trials Project[16]). In these latter expressions, however, GIS was part of what might otherwise be called digital history rather than spatial history because the approach was fundamentally archival and textual rather than driven by questions about space or even by geographical information.

Historical GIS is still a young sub-discipline. One of its leading advocates has defined it as having the "elements of *geohistoire*, historical geography, and spatial and digital history" and as being identified more by its characteristics than any theoretical approach or body of scholarship. Among these characteristics are the dominance of geographical questions and geographical information in framing inquiries, usually fashioned as patterns of change over time, and the use of maps to present its results.[17] But even though it is gaining use, especially among younger scholars, most historians—indeed, most humanists—have not adopted GIS or, more fundamentally, found it helpful. What remains puzzling to its practitioners is why the technology is not finding its way into the toolkit of these scholars. After all, human activity is about time and space, and GIS provides a way to manage, relate, and query events, as well as to visualize them, that should be attractive to researchers.

Significantly, the standard characterizations of historical GIS, as offered above, suggests the limits about the limits of GIS in history and the humanities, at least as currently practiced. GIS fundamentally is about what happens in geographic space. It relies heavily on quantitative information for its representations and analyses and views its results as geographical maps. There is no question that this calculus is valid and valuable, and it forces attention to important considerations, such as scale and proximity, that too often are absent from humanities scholarship. But it also is not the way humanists do their work. Quantitative humanists exist, of course, but the quantitative revolution forecast during the 1960s and 1970s as computers became less expensive and more powerful never materialized, or at least it never entered the mainstream of humanities scholarship. Humanists are drawn to questions and evidence that cannot be reduced easily to zeroes and ones. Yet the promise of GIS is so powerful—and the technology is becoming so ubiquitous—that we are loathe to abandon it too soon. Perhaps we have been asking the wrong question. Instead of musing about how we can get humanists to adopt GIS, it would be more fruitful to discover how to make GIS a helpmeet for humanists. Much of the work being done now fits neatly into what GIS was created to do. The real question is how do we as humanists make GIS do what it was not intended to do, namely, represent the world as culture and not simply mapped locations?

Currently, the problems with GIS as a platform for humanities research are well recognized. Spatial technologies in general, and especially GIS, are expensive, complex, and cumbersome, despite recent advances that have driven down costs and simplified the user experience. They require significant investments in time to learn both the language and techniques of the toolsets they employ. GIS and its cousins are literal technologies: they favor precise data that can be managed and parsed within a highly structured tabular database. Ambiguity, uncertainty, nuance, and uniqueness, all embedded in the evidence typically available to humanists, do not admit readily to such routinization. GIS also has difficulty managing time, which is a major problem in disciplines that orient their study to periods and epochs. Time is merely an attribute of space within a GIS, but it is a much more complicated concept for humanists, who well understand T.S. Eliot's sense of

> Time present and time past
> Are both perhaps present in time future
> And time future in time past.[18]

More important, the use of GIS requires humanists to be alert to issues that are not part of their training or culture. Humanists, for instance, are logo-centric. We find words, with their halos of meaning, better suited for describing the complexity, ambiguity, and uncertainty we see in our subjects, yet GIS relies heavily on visualization to display its results. It demands the use of spatial questions, whereas most humanists think rarely about geographical space and often do not understand how to frame a spatial query. It requires collaboration between technical and domain experts, thereby putting humanists, who work in isolation and are inept in the lingo, at a two-fold disadvantage. Finally, for many humanities scholars, GIS appears reductionist in its epistemology. It forces data into categories; it defines space in limited and literal ways instead of the metaphorical frames that are equally reflective of human experience; and, while managing complexity within its data structures, it too often simplifies its mapped results in ways that obscure rather than illuminate.[19]

Even if we were fluent in GIS, until recently the technology has had only limited ability to move us beyond a map of geographical space into a richer, more evocative world of imagery based on history and memory. But increasingly—and rapidly—it offers capabilities that we could employ with profit, although on the whole we have not. Over the past few years, GIScientists have made advances in spatial multimedia, in GIS-enabled Web services, geovisualization, cyber geography, and virtual reality that provide capabilities far exceeding the abilities of GIS on its own. This convergence of technologies has the potential to revolutionize the role of space and place in the humanities by allowing us to move far beyond the static map, to shift from two dimensions to multidimensional representations, to develop interactive systems, and to explore space and place dynamically—in effect, to create virtual worlds embodying what we know about space and place.

Seeking to fuse GIS with the humanities is challenging in the extreme, but already we have glimpses of what this technology can produce when applied to the problems in our disciplines. Within the field of cultural heritage, archaeologists have used GIS and computer animations to reconstruct the Roman Forum, for example, creating a 3-D world that allows users to walk through buildings that no longer exist, except as ruins. We can experience these spaces at various times of the day and seasons of the year. We see more clearly a structure's mass and how it clustered with other forms to mold a dense urban space. In this virtual environment we gain an immediate, intuitive feel for proximity and power. This constructed memory of a lost space helps us recapture a sense of place that informs and enriches our understanding of ancient Rome (Digital Roman

Forum Project).[20] In similar fashion, historians and material culturists have joined with archaeologists to fashion Virtual Jamestown. This project, in turn, is seedbed for an even more ambitious attempt to push the technology toward the humanities by placing Jamestown at one vertex of Atlantic World encounters. Its goal is to re-populate a virtual world with the sense of possibilities embedded in the past, what Paul Carter has called "intentional history."[21] Viewed within the spatial context for their actions, which includes the presence of proximate cultures, whether indigenous tribes, Spanish, Africans, or Dutch, we then can understand better how contingencies became lost as they butted against the encountered realities within the space the English claimed in 1607.

A paradigm project underway at West Virginia University aims to go even further by combining immersive technologies with GIS to recreate a sense of nineteenth-century Morgantown. Working from digitized Sanborn maps and extant photographs of buildings and streets, users enter a CAVE, a projection-based virtual reality system, and find themselves in another time and place, with the ability to navigate through an environment in which they now are a part. Soon they will be able to enter and explore a building, moving from room to room and examining the material objects within it. By adding sounds, smells, and touch, all within the capability of existing technology, this virtual reconstruction would engage four primary senses, making the experience even more real for participants. Once expensive, the costs of immersive environments are dropping rapidly, but, in fact, a CAVE is not essential for making an immersive environment open to humanists. As any parent of school-age children knows—or as any devotee of Second Life can testify—gaming technology already allows us to explore virtual worlds with a high degree both of verisimilitude and agency.

Even if it is becoming possible to imagine new, technology-based ways of exploring questions of heritage and culture, how do we make space, place, and memory dynamic and vital within them? With few exceptions, we have incorporated these elements into our Web sites and other digital products in much the same way we engage them in traditional scholarship, as part of an expert narrative. The primary evidence we use in each instance—documents, images, maps, material objects— represents personal and cultural memories that serve as mediators between us and the worlds they represent. We select and interpret these cultural artifacts to frame our understanding of the past and present. We use them within a book, an essay, or a Web site to structure a universe and make an argument. In this sense, technology makes more facile the process of knowledge creation we have always employed, but the difference we see most often is one of degree, not kind. We have not enabled our understanding of culture to be as dynamic as the act of creating culture itself, and it is to this end that we must direct technology if it is to help us open the past to the multiple perspectives and contingencies we know existed in the past.

The structuring of memory is especially problematic for GIS and other new technologies. Memory is essential for our identity, whether as individuals or as a society, but it remains troublesome as evidence because it always is informed by what has happened in the interim between an event and the act of recall. This condition makes memory dynamic, malleable, and contested. Except, perhaps, for intensely emotional events that remain fresh for us, we are remembering the last time we remembered. With each instance of recall, we remove even more of the contingency or sense of possibility that once existed. Through this process we construct the stories of ourselves, and in this way we create the various narratives that recount our communal history. But unlike personal memory, which seeks to reconcile or hide our interior conflicts, communal memory becomes contested public space. The stakes of this struggle are high because the outcome confers legitimacy, yet we also know that memory privileges what we want or need to believe. As a society, it means that we have often removed from our public memory the voices of dissent, and we have expunged from our physical and cultural landscape the "shadowed ground" that reflects our shame.[22]

How then do we attempt to recover the unrecoverable and find our way through memory to identity and culture? Of course, we cannot, and it is futile to try. We live only in the moment poised precariously between past and future, conscious of the influence of both. But what we can do is inform the present more fully with the artifacts of social memory, the evidence of recall from various times and various perspectives. One means to this end is through "deep mapping," an avant-garde technique first urged by the Situationist International in 1950s France. Popularized by author William Least Heat-Moon in *PrairyErth (a deep map)*,[23] the approach "attempts to record and represent the grain and patina of place through juxtapositions and interpenetrations of the historical and the contemporary, the political and the poetic, the discursive and the sensual...."[24] In its methods deep mapping conflates oral testimony, anthology, memoir, biography, images, natural history and everything you might ever want to say about a place, resulting in an eclectic work akin to eighteenth and early nineteenth-century gazetteers and travel accounts. Its best form results in a subtle and multilayered view of a small area of the earth.

Described as a new creative space, deep maps have several qualities well-suited to a fresh conceptualization of humanities GIS. They are meant to be visual, time-based, and structurally open. They are genuinely multimedia and multilayered. They do not seek authority or objectivity but involve negotiation between insiders and outsiders, experts and contributors, over what is represented and how. Framed as a conversation and not a statement, deep maps are inherently unstable, continually unfolding and changing in response to new data, new perspectives, and new insights.

It is not necessary to adhere to hazy theories of psychogeography or to the neo-Romanticism of the British idea of "spirit of place" to find an analog between the deep map and advanced spatial technologies. Geographic information systems operate as a series of layers, each representing a different theme and tied to a specific location on planet earth. These layers are transparent, although the user can make any layer or combination of layers opaque while leaving others visible. In the environmental sciences, for example, one layer might be rivers and streams, another wetlands, a third floodplains, a fourth population, a fifth roads and bridges, a sixth utility lines, and so forth. By using information about rainfall amounts and rates within a predictive model, we can turn on and off layers to see what areas and which populations, habitats, and infrastructure will be affected most quickly by flooding and how best to plan for relief and recovery. We can view these layers in the sequence predicted by the model or we can view only the layers that most immediately affect human health and safety.

A deep map of heritage and culture, centered on memory and place, ideally would work in a similar fashion. Each artifact—a letter, memoir, photograph, painting, oral account, video, and so forth—would constitute a separate record anchored in time and space, thus allowing us to keep them in relationship, and each layer would contain the unique view over time—the dynamic memory—of an individual or a social unit. The layers could incorporate active and passive cultural artifacts, such as memories generated by intentional recall as well as memories left to us in some fixed or material form. They also might contain accounts from the natural world, such as found in meteorological and geological records. The layers of a deep map need not be restricted to a known or discoverable documentary record but could be opened, wiki-like, to anyone with a memory or artifact to contribute. However structured, these layers would operate as do other layers within a GIS, viewed individually or collectively as a whole or within groups, but all tied to time and space as perspectives on the places that interest us.

The deep map is meant to be visual and experiential, immersing users in a virtual world in which uncertainty, ambiguity, and contingency are ever-present, influenced by what was known (or believed) about the past and what was hoped for or feared in the future. It is here that traditional GIS faces its sternest test: it cannot yet create such a rich visual environment, much less work with such imprecision and fluidity as the nature of humanities questions and evidence demands. But the rapid convergence of GIS with other technologies, especially multimedia and gaming tools, suggests that we are not far from the point when it will be possible to construct deep maps and landscapes of culture for any place where people leave records of their experiences.[25]

When this happens, what will it mean for us as humanists? Assuming continued progress in making the technology more complete and easier to use, it is possible to construct at least two views of a GIS-based landscape of culture and place. In the first scenario, humanities GIS is a powerful tool in the management and analysis of evidence, contributing primarily by locating historical and cultural exegesis more explicitly in space and time. It aids but does not replace expert narrative: it finds patterns, facilitates comparisons, enhances perspective, and illustrates data, among other benefits, but its results ultimately find expression primarily in the vetted forms accepted by our disciplines. In this view, GIS provides geographical context and depth to an expert interpretation of the past. It represents, at heart, a maturing of our current use of GIS.

In the second scenario, the technology offers the potential for an open, unique postmodern scholarship, an alternate construction of history and culture that embraces multiplicity, simultaneity, complexity, and subjectivity. Postmodernist scholarship has sharply challenged the concept of objectivity in history, which has been the lodestar of so-called scientific history since the late nineteenth century. It rejects the supremacy of empiricism, an Enlightenment concept, in favor of knowledge based on all the senses. Postmodernism also has called into question the primacy of texts and logic as the foundation of knowledge. In its epistemology, history is not a grand narrative—an authoritative story of a society's past—but instead a fragmented, provisional, contingent understanding framed by multiple voices and multiple stories, mini-narratives of small events and practices, each conditioned by the unique experiences and local cultures that gave rise to them.

A humanities GIS-facilitated understanding of society and culture may ultimately make its contribution in this way, by embracing a new, reflexive epistemology that integrates the multiple voices, views, and memories of our past, allowing them to be seen and examined at various scales; by creating the simultaneous context that we accept as real but unobtainable by words alone; by reducing the distance between the observer and the observed; by permitting the past to be as dynamic and contingent as the present. In sum, it promises an alternate view of history and culture through the dynamic representation of memory and place, a view that is visual and experiential, fusing qualitative and quantitative data within real and conceptual space. It stands alongside—but does not replace—traditional interpretive narratives, and it invites participation by the naïve and knowledgeable alike. We are not yet at this point, but some day we could be. It is a vision worth pursuing.

NOTES

1. Doreen Massey, *For Space* (London: Sage Publications Ltd., 2005).
2. Brian Jarvis, *Postmodern Cartographies: The Geographical Imagination in Contemporary American Culture* (New York: Palgrave Macmillan, 1998), 1–6.
3. David N. Livingstone, "Science, Region, and Religion: The Reception of Darwin in Princeton, Belfast, and Edinburgh," in Ronald Numbers and John Stenhouse, eds., *Disseminating Darwinism: The Role of Place, Race, Religion, and Gender* (Cambridge: Cambridge University Press, 1999), 7–38. Cited here at 7.
4. Jarvis, *Postmodern Cartographies*, 7–8.
5. John Coppock and David Rhind, "The History of GIS," in David Maguire, Michael Goodchild, and David Rhind, eds., *Geographical Information Systems: Principles and Applications. Volume I: Principles* (London: Longman Scientific and Technical, 1991), 21–43.

6. Timothy Foresman, "GIS Early Years and the Threads of Evolution," in Timothy Foresman, ed., *The History of Geographic Information Systems: Perspectives from the Pioneers* (Upper Saddle River, N.J.: Prentice Hall, 1998), 3–17.

7. John Pickles, ed., *Ground Truth: The Social Implications of Geographic Information Systems* (New York: Guildford Press, 1995).

8. Nadine Schuurman, "Trouble in the Heartland: GIS and Its Critics in the 1990s," *Progress in Human Geography* 24 (2000), 569–90.

9. Nadine Schuurman, *GIS: A Short Introduction* (Oxford: Wiley-Blackwell, 2004).

10. Eric Sheppard, "Knowledge Production through Critical GIS: Genealogy and Prospects," *Cartographica* 40 (2005), 5–21.

11. Robert Rundstrom, "GIS, Indigenous Peoples, and Epistemological Diversity," *Cartography and Geographic Information Systems* 22 (1995), 45–57.

12. Merrick Lex Berman, "Boundaries or Networks in Historical GIS: Concepts of Measuring Space and Administrative Boundaries in Chinese History," *Historical Geography* 33 (2005), 118–33.

13. Geoff Cunfer, *On the Great Plains: Agriculture and Environment* (College Station: Texas A&M University Press, 2005).

14. Michael McCormick, *Origins of the European Economy: Communications and Commerce, AD 300–900* (Cambridge: Cambridge University Press, 2001).

15. Valley of the Shadow: Two Communities in the American Civil War: http://valley.vcdh.virginia.edu/ (accessed 9 Jan. 2009).

16. Salem Witch Trials: Documentary Archive and Transcription: http://etext.virginia.edu/salem/witchcraft/ (accessed 9 Jan. 2009).

17. A. Knowles, "GIS and History," in A. Knowles, ed., *Mapping the Past: How Maps, Spatial Data, and GIS Are Changing Historical Scholarship* (Redlands, Calif.: ESRI Press, 2008), 1–26.

18. T.S. Eliot, "Burnt Norton," in T.S. Eliot, *Complete Poems and Plays, 1909–1950* (New York: Harcourt, 1952).

19. David J. Bodenhamer, "History and GIS: Implications for the Discipline," in A. Knowles, ed., *Mapping the Past: How Maps, Spatial Data, and GIS are Changing Historical Scholarship* (Redlands, Calif.: ESRI Press, 2008), 220–33.

20. Digital Roman Forum project: http://dlib.etc.ucla.edu/projects/Forum (accessed 9 Jan. 2009).

21. Paul Carter, *The Road to Botany Bay: An Exploration of Landscape* (London: Faber and Faber, 1987), 3.

22. See, for example, Kenneth E. Foote, *Shadowed Ground: America's Landscape of Violence and Tragedy*, rev. ed. (Austin: University of Texas Press, 2003); David Lowenthal, "Past Time, Present Place: Landscape and Memory," *The Geographical Review* 65:1 (1975), 1–36.

23. William Least Heat-Moon, *PrairyErth (a deep map)* (Boston: Houghton Mifflin Company, 1993).

24. Michael Pearson and Michael Shanks, *Theatre/Archeaology: Disciplinary Dialogues* (London: Routledge, 2001), 64–65.

25. David Bodenhamer, "Creating a Landscape of Memory: The Potential for Humanities GIS," *International Journal of Humanities and Arts Computing* 1:2 (2007), 97–110.

THREE

Geographic Information Science and Spatial Analysis for the Humanities

KAREN K. KEMP

INTRODUCTION

Geographic Information Science (GISci) is the science behind the technologies of Geographic Information Systems (GIS). As a science, GISci evolved in a context of precision, quantitative measurement, and notions of accuracy. As such, it might seem that its technology has little application in the humanities where imprecision, qualitative information and individual, sometimes conflicting, interpretations of "facts" are the norm. Fortunately, GISci has a strong intellectual foundation in the discipline of geography, a field that sits astride the science/social science divide, and its practitioners are generally comfortable addressing the challenging issues that arise when we attempt to represent the complex and ever changing places in which we live within the rigorous structure of the digital computer.

In a somewhat circular definition, GISci is an information science that focuses on the collection, modeling, management, display, and interpretation of geographic information. It is an integrative field, combining concepts, theories, and techniques from a wide range of disciplines, allowing new insights and innovative synergies for increased understanding of our world. By incorporating spatial location (geography) as an essential characteristic of what we seek to understand in the natural and built environment, geographic information science and systems GIS provide the conceptual foundation and synergistic tools to explore this frontier.[1]

Geographic information is defined simply as any data or information that has a geographic reference. While attaching latitude and longitude coordinates to a piece of data does create geographic data, there are many other ways to assign location, which are discussed later. It may be useful to note that GISci authors seem to spend an inordinate amount of printed text distinguishing data, which are facts, from information, which includes some interpretation, though one frequently finds that one person's information is another person's data. In this chapter, an attempt is made to keep to the facts versus interpretation distinction while not being overly strict in the definitions.

What is important about assigning a geographic reference to data is that it then becomes possible to compare that characteristic, event, phenomenon, etc. with others that exist or have existed in the same geographic space. What were previously seemingly unrelated facts become integrated and correlated. Importantly, it allows us to perform spatial analysis, which might be thought of simply as what we do with geographic information once it is in the computer. Spatial analysis helps us understand what is going on in our geographic information. As will be seen later, however, the term has been defined in so many different ways that a clear practical definition is difficult to constrain.

For humanists who wish to begin seeing their work in a spatial perspective, perhaps even to work with GIS, it is necessary to understand some of the key fundamental themes in geographic information science that are essential in modeling and analyzing the world using a computer. As will become clear, many of the basic assumptions on which the technology was designed do not play well with the methods and information used in the humanities. Fortunately, this misalignment is the point at which insight and learning can take place, making the application of spatial reasoning and geographic technologies a new frontier in the humanities.

The chapter begins with a discussion of how we conceptualize, categorize, and represent the geographic complexity of the world. Where and what we describe about the places and people we are interested in are key characteristics we must articulate clearly. Once we have spatially conceptualized the things of interest, we must create representations of these conceptualizations in geometric form so they can be quantified and stored as bits of data in the computer. Once the data is stored, the fun begins. Then we can start analyzing our data and turning it into information and hopefully, with insights gained, into knowledge. All of these stages are required, whether we are trying to understand physical process in the natural environment or human processes across the social, cultural, and historical landscape.

DECOMPOSING THE INFINITELY COMPLEX WORLD

The world is composed of an infinite number of things and characteristics. Look at a landscape and you might see lakes and mountains, or you might see trees and meadows, or maybe the roads and rivers. What you see is determined by what you are looking for or what interests you. If you ask a group of people to individually sketch a map of "here," each person will create a different map composed of different things. What you see or record on a sketch map depends on your professional background and interests, the size and nature of the study area, the tools you have available to measure phenomena in the environment, the data that you already know exists, and so on. Let us consider a few of these key aspects.

SCALE

Scale is perhaps one of the most important characteristics geographers and others consider, consciously or unconsciously, when they are trying to describe what they see in the world. Unfortunately, the definition of the term is often muddled. One way we often use the term scale is to describe the size of the region we are considering. Thus a large-scale project might mean one that involves a large region, a lot of people and/or a lot of money. A small-scale project will just cover a small area and/or cost a small amount of money.

In contrast, in geography and thus in GISci, there is a technical definition of the term scale that has a completely opposite meaning. In this formal definition, scale is measured as the mathematical relationship between the real world and its representation, usually on a map. So, scale may be, for example, expressed as "1 cm on the map represents 1000 km on the ground," or, perhaps, a map may be said to be "1 inch to the mile." Formally this relationship is called the representative fraction and often is stated as a ratio; the scale 1:50,000 means 1 unit of length on the map represents 50,000 units on the ground. When speaking of scale in this manner, then, a small-scale map is one that has a small ratio with the right

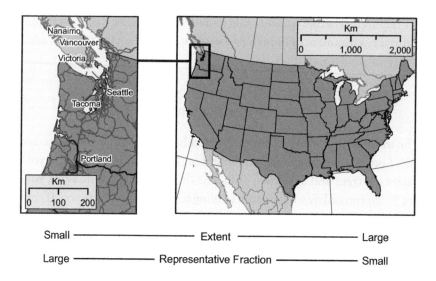

FIGURE 3.1. Two ways of thinking about small scale versus large scale.

side being a large number, such as 1:1,000,000. This map would show a very large area and, using the first definition, it would be a large scale map (see Figure 3.1). Hence, the confusion.

To the purist, only the second definition is correct, but in the rapidly democratizing world of GIS—think Google Earth and other sites on the Web that allow everyone to become cartographers—we have to accept and acknowledge both. Suffice it to say, whenever anyone mentions scale in the context of GIS, be sure that the meaning is agreed upon by all parties. We will return to exploring the role that scale plays in GIS later in this chapter.

No matter the definition, scale is essential in understanding how we conceptualize the world. If we consider the world on a continental scale, the entities of interest will be broadly defined and their locations perhaps imprecise though still capable of being mapped. If our area of interest is very local, then there will be many detailed items in our world and their locations may be very precisely known, or not. Importantly, if we are thinking about integrating studies that have both regional and local perspectives, the difference in how we view the world at these different scales will have profound impacts on what we record and analyze.

OBJECTS OR FIELDS?

When we look out across a landscape, we see the world filled with objects scattered across a continuous background. Think about a pastoral landscape. The background of rolling green fields is dotted with objects—houses, barns, fences, cows, roads, and farmers on tractors. In 1992, in what is now a classic article in the relatively young field of geographic information science, Helen Couclelis, who often writes classically styled philosophical essays in GISci, posited that "people manipulate objects (but cultivate fields)."[2] This statement encapsulates an important tension that we confront as we begin thinking about representing the complexity of our places in the computer.

Some of the entities in the world are continuous. The land surface is perhaps the easiest to think of in this manner, but there are many other such entities, including air temperature or pressure, noise levels, and soil moisture. In the continuous view of the world, it is possible to measure the value of the phenomenon we are interested in at any location. You can go anywhere on land and measure a value of elevation, and you can measure the value at almost infinitely fine scales—once every kilometer, once every meter, centimeter, and so forth—until you reach the molecular level. In GIS, we do not usually go to such small scales, but we do agree that the land surface is a continuous phenomenon. In physics, continuous entities are called "fields," and this term is often used in GIS. In a spatial field, the value of the phenomenon under study is associated with the location at which it is measured and a value can be measured anywhere.

We conceive other entities in the world as discrete objects. In fact, if you ask someone to describe a scene or a landscape, most of their description will involve objects. Objects often, though not always, have discrete boundaries and may be moveable. Objects are scattered across our areas of study and, depending on what objects of interest we have defined, there are likely to be many locations where there are none of these objects. In our pastoral landscape, for example, between the cows there are no cows. At any point you could measure "cow" and the result would be simply yes or no. Importantly, objects have identities that are usually unrelated to their location. Thus the cow Daisy exists no matter where in the pasture it grazes.

Of course, it is not quite so simple. A lot of objects do not have discrete boundaries, mountains being an excellent example. Conceptually, a named mountain is an object. You can point to it on a map at a specific location and elsewhere that mountain does not exist. But where is the edge of the mountain? Where does it begin as you move from a location that is not that mountain to a location that is mountain? There is no absolute answer to this question; its implementation is one of the largely unresolved challenges in most geographic information systems and one which has major implications in the spatial humanities. Here many of our objects of interest are spatially imprecise or uncertain, often as a result of gaps in the historical record or due to different interpretations of the object. For example, in the U.S., neighboring Native American tribes and the federal government may all fix the boundaries of Indian territories in different locations.

An advanced approach to handling this challenge, rarely implemented, involves the formal use of fuzzy set theory. Simply, rather than a location being either mountain or not mountain, fuzzy set theory allows us to measure the "mountainness" of each location. At the peak of the mountain it is definitely mountain, for instance, 5 on a scale of 5. In the valley it is definitely not mountain, 0 on the scale. In between, where the land just begins to slope upward, its "mountainness" might register as 2. This measure of belonging to a class allows users to impose various interpretations of where class boundaries lie, and it supports advanced mathematical analyses.

Sometimes we create pseudo-fields from objects. For example, the map in Figure 3.2 shows population density of the U.S. in 1850. The shaded areas cover the map continuously (except for that gap of the still "unorganized" territory) so it is possible to determine a value at any location. However, these maps are actually based on aggregate counts of the number of individuals (objects) within defined areas, not values of a continuous field. Importantly, the values are dependent completely upon the size and shape of the regions over which individuals are aggregated. Figure 3.3 illustrates this problem known as the Modifiable Areal Unit Problem (MAUP). The MAUP is another of the great unresolved and often overlooked challenges in GIS. It is particularly important when we use historical census data because the areas over which populations are counted

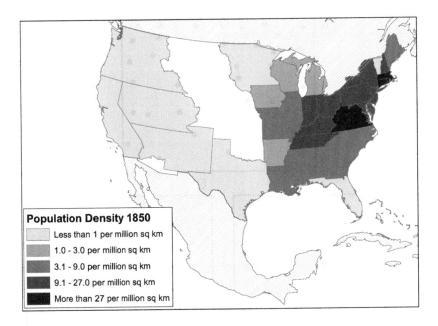

FIGURE 3.2. Map of population density. This historic census data and boundary files were downloaded from the National Historical Geographic Information System at www.nhgis.org.

rarely have the same boundaries over time. How can we estimate changes in populations when the numbers change as a result of area changes? Do they reflect real changes in the population characteristics or just different arrangements of people?

CATEGORIES, ONTOLOGIES, AND SEMANTICS

Another question we think about when we try to articulate what we will put in a GIS is: what are the things we are going to "map" or store in our database? Humans have a natural inclination to categorize what we see in the world. It helps us build structure in the otherwise overwhelming complexity our senses perceive. When we look at a crowd, we see "people." But what we see depends upon how we think about the world. For instance, when most people look at a forest closely, they see "trees," "shrubs," or "grasses," but when botanists look at a forest, they see various species of plants, all with precise definitions and names.

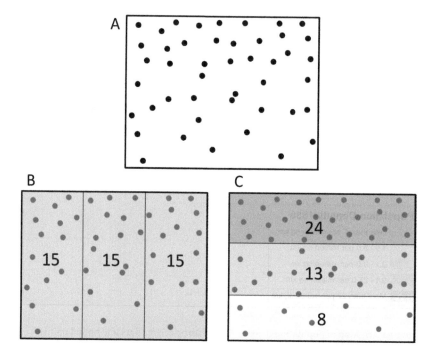

FIGURE 3.3. An example of the modifable areal unit problem. Box A shows a distribution of entities across a landscape. Boxes B and C show the same distribution grouped into three areas, the numbers show how many entities are in each area. The shading reflects the relative density of entities within each region. Note how the resulting density map would change based on changes in the area boundaries.

From categories we build ontologies, a term that is gaining increasingly prominent use in GISci as our databases become large and data is shared among them. This term has both philosophical and computing foundations. In philosophy, there is only one ontology, the single, fundamental structure of how humans understand the world. Ontology is simply the "study of being" or the "study of what is." This definition is not pragmatic enough for computer scientists, of course, so in the computing context an ontology is a description of the concepts in an area of knowledge. Such descriptions identify the kinds of entities that exist and their relationships. Unlike in philosophy, a multitude of ontologies can and do co-exist. Thus, as we seek to build computer representations of our complex world by identifying the objects and fields that interest us,

GIS AND SPATIAL ANALYSIS FOR THE HUMANITIES · 39

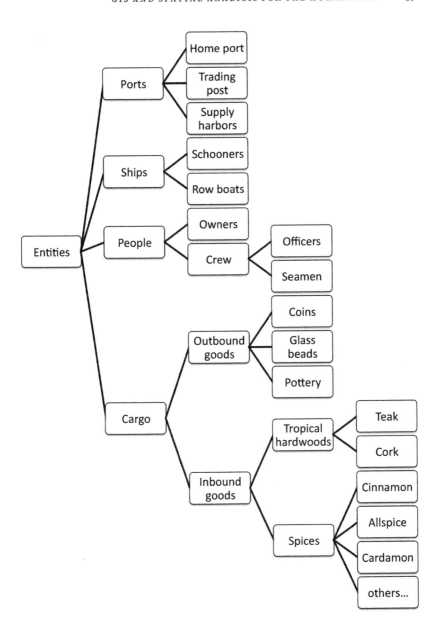

FIGURE 3.4. A simple ontology for the spice trade in the sixteenth century.

we identify the categories, how they are distinguished from one another, how one category is related to other categories, and if there are subcategories. Ontologies are generally visualized as tree structures with links between branches. Figure 3.4 depicts a simple ontology we might build for the spice trade of the sixteenth century.

Ontologies are the fundamental means by which we design our databases for GIS. In the world of database management, this stage is called database design and in it we construct what is called the database schema, which essentially is an ontology with all the characteristics of and relationships between each kind of entity in the database specified. For smaller projects, often in humanities GIS projects, a formal ontology may not be constructed at the beginning; rather, it evolves organically from the data collection efforts. While this approach may require a complete redesign of the system later, sometimes it is useful to proceed slowly, gradually constructing a structure for understanding the components of the world we are studying.

Finally, it is important to mention semantics, the meaning of the words we use to name categories and thus the terms in our ontologies. While all of us might agree on the name of a category being "road," wildlife ecologists may conceptualize it as a boundary that is a barrier to animal movement, economists as a line that forms part of a transportation and communication network, and engineers as an area of pavement that has a specific width, thickness, and surface slope. Trying to incorporate all these meanings into a single database schema would be impossible. Thus we build different ontologies for different communities and efforts may be made later to "cross-walk" the ontologies, identifying terms with different words for similar meetings and vice versa. While a discussion of semantics in GISci can fill books[3] we mention it here simply as a potential area of confusion about which all users of GIS should be aware.

ATTRIBUTES

Briefly, it is important to mention the term attribute. In GIS, this is such a commonly used and understood concept that its practitioners are often surprised to discover others do not understand it. Attributes simply are the characteristics that we identify and record about the entities in our database. So a collection of information about "towns" might include the

TABLE 3.1. Part of an Attribute Table for a GIS Layer of Towns

Place Name	County	Grid Reference	Year Established	Population 1850	Population 1860
Abingworth	Sussex	TQ1016	1771	280	269
Ablington	Wiltshire	SU1546	1475	1830	1848
Abney	Derbyshire	SK1979	1417	544	571
Aboyne	Aberdeenshire	NO5298	1412	1180	1286
Abram Brow	Lancashire	SD6101	1676	390	363
Abriachan	Inverness-shire	NH5535	1357	2990	2930

attributes of name, population, date founded, and location. Attributes are often structured into tables, conveniently called attribute tables. Table 3.1 shows a small part of an attribute table with each row containing information about a single entity and columns containing information about a single attribute. Note, these columns are often called "fields," but this should not be confused with the different use of this term as it is described above.

There are generally considered to be five different categories of values we can store as attributes. While the same alphabetic or numeric characters may be used to express attribute values for various categories, understanding these semantic and functional differences is critical. These categories are:

- Nominal—alphanumeric values that are labels, names, or IDs. They have no mathematical value. Examples are people's names or social security numbers, archaeological site numbers.
- Ordinal—alphanumeric values that have an inherent order, but the "distance" between adjacent values cannot be calculated. Examples are places in a race or competition (first place, second place, last place) or suitability ranking (best, good, poor).
- Interval—numeric values that are a quantitative measure, but the scale on which they are measured has no absolute zero. Years and elevation are two good examples. The number we assign to a year depends on when we choose to begin counting years, and it is possible to have "negative years" (BCE). Likewise, elevation is measured from sea level, but the precise positioning of sea level varies even within single countries, and it is quite possible to have negative values (below sea level).

- Ratio—numeric values that measure quantity; there is an absolute zero and negative numbers are not possible. Examples include a person's height, amount of rainfall, number of years a site was occupied.
- Cyclic—numeric values that have a limited top value and that cycle back to 0 after that value is reached. Degrees on a compass are an example of this. Continue clockwise from 359 degrees and the next number is 0 degrees.

Understanding these categories is important in ensuring that inappropriate operations are not performed on data. Nominal and ordinal values cannot be manipulated by mathematics. For example, while the number 10 may be stored in the computer as a value for any of these categories, if it is an ID (a nominal value), we cannot subtract it from ID number 15 to get any meaningful result. If it is the rank of longevity of emperors, multiplying it by 2 would likewise produce a meaningless number. Interval numbers can be subtracted (year 1995 minus year 1965 gives 30 years) but they cannot be divided or compared as ratios (year 1000 is not half of year 2000). Cyclic numbers have problems with addition and subtraction when the value will pass the 0 point (345 degrees plus 50 degrees is 35 degrees). Keeping these distinctions in mind when coding data into a database is important for how it might be manipulated later.

DETERMINING AND SPECIFYING LOCATION

Finally, we come to location. In order to build a GIS, which by definition requires "geographic information," we must have some means of formally describing the location of a place. First, location can be relative or absolute. The former describes locations relative to the location of other places. So, northwest of London, beside the lake, and 100 miles east of the crossroads are simple relative locations. It is possible to build a map using only relative locations, if we begin by arbitrarily assigning a position for the one entity to which all others are directly or indirectly related. So, put the crossroads at the center and all other related descriptions can be mapped.

However, relative location only works for a single set of related data. If we want to see how very different sets of data are related in space, then we need to find some way of anchoring them to the earth so that interactions in geographic space can be explored. This is absolute location. Absolute location fundamentally requires at least some of the data to be anchored to a coordinate system. Latitude/longitude, often shortened to lat/long, is a universal coordinate system, widely used, but there are many others including Universal Transverse Mercator (UTM), State Plane, and various National Grids. All of these horizontal coordinate systems anchor point locations to the earth's surface, and they are described exhaustively in any text on cartography or GIS. Some of the modern coordinate systems also allow a vertical coordinate to be designated, so that it is possible to specify a location in 3-D space, above or below the earth's surface.

Once we begin to work with coordinate systems, the advanced concept of datums comes into play. The accurate and precise specification of the location of points on the earth's surface is the domain of the field of geodesy. With the huge advances in global positioning systems that allow us to determine location from a perspective in space as opposed to a perspective tied to the place we are trying to measure, location measurements can now be so accurate and precise that it is possible to record the changing latitude and longitude of locations as the surface of the earth moves under the forces of plate tectonics. At this level of precision, the lat/long values of places on the earth's surface actually depend on date and time. But for most of us, this kind of precision and accuracy is unnecessary.

You can think of a geodetic datum as a spherical grid that is laid over the earth that we use to measure the latitude and longitude values for any location. We could orient and shape this grid in an infinite number of ways. Fortunately, the World Geodetic System 1984 (WGS 84) is now a global standard. It and the virtually equivalent North American Datum 83 (NAD 83) are the datums you should use as much as possible when working with geographic coordinates. Most GISs have translation functions to recalculate latitude and longitude coordinates from other datums. Given a consistent datum, we can assume that a location on the earth's surface will always have the same latitude and longitude.

A related dichotomy in how we specify location is that of direct and indirect location. Direct locations are those that are stated in coordinates as described above. Indirect location uses references to other objects whose direct location is known. The two most commonly used indirect

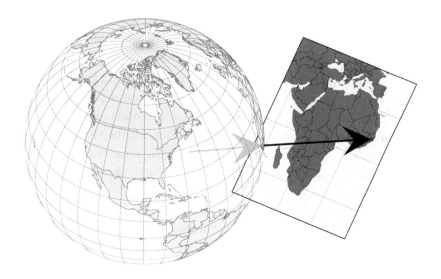

FIGURE 3.5. Making a projection of Africa (*reversed*) through an Earth "wireframe."

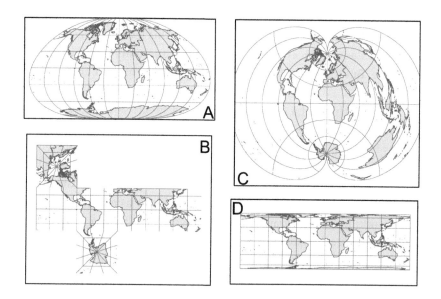

FIGURE 3.6. A few different map projections. A) Loximuthal. B) Cube (this projection is designed to be cut out of paper and folded into a cube). C) Polyconic. D) Cylindrical equal area.

references are place names and addresses. If you have a GIS dataset (i.e., a GIS map) of cities and a separate attribute table that contains the names and other information about a subset of those cities, you can associate that attribute table with geographic location through the common key of "name." This capability allows you to explore and map all of these associated attributes in the spatial context of the GIS, though you started only with a simple table of text values. Addresses work in a similar way if there is a GIS dataset that contains the locations of the streets and addresses along those streets. The entities described by any attribute table with an address as one of the attributes can then be associated with this geography.

Indirect locations are vitally important in the spatial humanities. Importantly, in order to use a geographic reference that is not a coordinate, it is necessary to have or to construct the geographic framework of the named locations. Gazetteers are one major source of this kind of framework, but there are others. Constructing this geography can be a significant challenge when trying to map historic events from text documents. Ruth Mostern, a historian who has written extensively on the design of gazetteers as a foundation for historical GIS, has spent several years mapping the changing geography of administrative units during the Song dynasty in China from geographic descriptions recorded in administrative documents from that period.[4]

PROJECTIONS

Another advanced concept worth mentioning here is projection, which is the method by which the spherical earth is "projected" onto a flat surface (i.e., a sheet of paper or a computer screen). Think of peeling an orange (a sphere) and pressing out the peel so that it is flat (like a piece of paper). Clearly, you cannot make the peel rectangular simply by pressing it down.

Figure 3.5 shows one kind of projection. While many projections are geometrical operations like this, many of them require mathematical equations to convert latitude and longitude coordinates into rectangular map coordinates. The important point to remember is that maps made with different projections cannot be overlaid—the same location will appear in different map locations if the projections are different. Figure 3.6 illustrates this.

SPATIAL AUTOCORRELATION

One final term, spatial autocorrelation, must be introduced as it is upon this concept that all of GIS (and much of geography) is founded. Spatial autocorrelation is famously described by computer cartography pioneer Waldo Tobler's First Law of Geography, "Everything is related to everything else, but near things are more related than distant things."[5] It is because of spatial autocorrelation that we can construct digital representations of our infinitely complex world. We cannot possibly categorize, code, and record everything in the world, but because of spatial autocorrelation, we can make many good guesses about the places not mentioned in our databases that are near to the places included in them. The elevation of the land surface is a good example of this. We can measure and store in our database the elevation at a number of points and know that locations close to these points are likely to have similar elevations. Spatial autocorrelation does break down in certain contexts (changes in taxation rates across a national boundary, for example) but wherever there are distinct boundaries, it is useful to keep in mind that people and places each side of that boundary may be more similar to each other than to places farther away (local dialects for example).

REPRESENTING GEOGRAPHIC CONCEPTS DIGITALLY

Now that we have conceptualized the world in formal structures that involve ontologies, categories, objects, fields, and scale, we have to put all of that into the computer. The next step is to decide how to represent the geography of our formalized world. The field/object dichotomy gives us two fundamental data models for GIS—vectors and rasters. How we represent the world determines how we can analyze it, so it is important to understand this important divide in GIS.

Objects are usually represented in GIS in the geometric form of points, lines, and polygons (areas). Formally, this means that the location of a point (a 0-dimension object) is specified by a pair of coordinates, lines (1-dimension) are a connected series of points and polygons (2-dimensions) are defined by a connected set of lines. Current GIS are not well equipped to handle 3-dimensional objects, so we will not explore

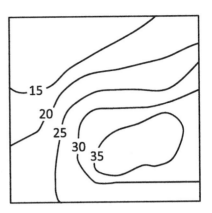

FIGURE 3.7. Elevation depicted as a raster (*left*) and as contour lines which are vectors (*right*).

that here. Collectively this form of representation is called vectors. Like the objects, vectors define precise locations and sharp boundaries. The topographic maps that most of us are familiar with are vector maps.

An important aspect of the vector data model is its ability to store and allow the analysis of geometric relationships, which is usually called topology, a concept that originates in mathematics. The development in the late 1970s of simple relational database structures that can store topology spurred the evolution of GIS and spatial analysis as we know it today. By storing (or determining on-demand) topology, the computer can quickly determine what polygons are adjacent, what points fall inside a polygon boundary, or which line segments connect to others in a network.

Since fields are continuous, the computer representation for them also must be continuous in concept. The most commonly used representation for fields are rasters. Here, the surface of the earth is divided into a grid of (usually) rectangular cells. Each cell is assigned a value representing the average (or sometimes maximum or minimum) measurement of the characteristic of interest over that area on the earth's surface. Satellite images are common examples of rasters. If you zoom into an image far enough, the rectangular cells become visible. The values stored for each cell is often displayed as ranges (e.g., less than 10, 10–20, 21–30, etc.) and assigned specific color hues or intensities. Figure 3.7 shows elevation mapped as vectors and as a raster.

Once again, it is not quite that simple. It is possible to represent objects as rasters (think about pictures you take with your digital camera—you can discern the people in the images, even though it is raster data) and fields as vectors (here the best example is the field of elevation that we commonly represent on topographic maps using contour lines, as shown in Figure 3.7). Generally, however, it is useful to think of vector representations of objects and raster representations of fields. If that does not match the conceptualization, then here is one of those important places of disconnection from which additional insight can come.

It is important to mention briefly another aspect about scale that comes into play once we represent our data digitally. Earlier, we mentioned the problem of integrating data that is conceptualized on a regional scale with data that is conceptualized on a local scale, in which case the ontologies may differ. Additionally, if the same feature is shown on maps of very different scales, there will always be a difference in their representations when it is extracted and stored in GIS. A line captured from a larger scale map is likely to be more sinuous than from the small scale map. Points may appear in different locations due to the different levels of precision with which location can be determined. So, before GIS (or map) data from very different sources can be used together, it is essential to resolve differences in how similar objects are represented.

SPATIAL ANALYSIS FROM A TO Z

Spatial analysis can be defined very broadly as any form of data analysis in which the results depend on location and will change if the location of the objects under study changes. In a GIS, once we have the geographic and related attributes stored in the computer, spatial analysis can begin. In its simplest form, spatial analysis takes place when we look at a map. We see relationships between entities (suggesting causal relationships), concentrations of objects ("hot spots"), and variations across the landscape that inspire understanding. Of course, this is not really simple as these kinds of insights from visual analysis result from the complex interaction of our innate visual and computational capacities. These human abilities are so powerful that many of the mathematical and geometric techniques used in computational spatial analysis have been designed to mimic them.

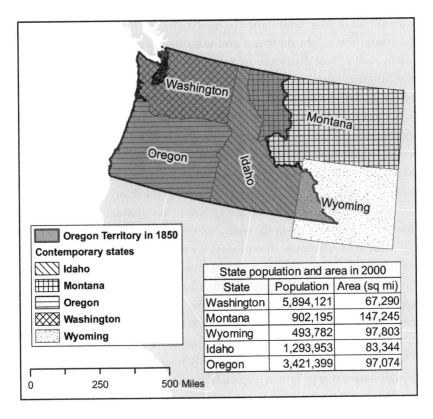

FIGURE 3.8. An example of areal interpolation. If Oregon Territory had a population of 13,294 in 1850 and an area of 288,798 sq. mi., can we determine the population change in the Idaho region in the last 150 years? One answer is Idaho falls entirely within Oregon Territory and its area is 18.5% of the total (83,344/288,798 × 100), so we can estimate it had 2,459 people in 1850 and thus a population increase of over 80,000 people in 150 years.

Beyond visual analysis, there is a plethora of spatial analysis techniques. Fortunately, in the humanities, as elsewhere, the handful of techniques most often used are conceptually uncomplicated. While it is not possible in a brief chapter to provide a summary of all relevant techniques, the following "A to Z sample" illustrates a few of the most commonly used ones. Many others are illustrated elsewhere in this volume.

Areal interpolation allows data aggregated to a particular set of zone boundaries (population counts for census tracts) to be mapped onto a different set of zone boundaries (school districts). It is applied to vector data only. You might do this if you want to know how many students live in each school district or to determine population change between two different population censuses (there are always some census tract boundaries that change between censuses). In the most simple form of areal interpolation, it is achieved by determining the proportion of each source zone covered by each target zone (see Figure 3.8). This proportion is used to apportion the total count of the source zone to each overlapping target zone so that the count in the target zone is a sum of all the portions calculated for each overlapped source zone.[6] Like many spatial analysis techniques, this version of areal interpolation has a fundamental assumption that the population in the source zones are evenly distributed across the study area. Of course, few objects are evenly spread across space, so here the modifiable areal unit problem mentioned earlier comes into play. How do we account for unequal distributions and boundaries that change over time? Fortunately, many enhancements and refinements are available to the simple procedure, and these techniques, too numerous, and too complex to be described simply, will be useful to humanists.

Buffer is perhaps the most frequently used form of spatial analysis, though some experts claim it is spatial data manipulation, not spatial analysis. While it is implemented in very different ways for the two data models, raster and vector, conceptually it is very simple—a zone is extended beyond the object of interest to a specified distance. Figure 3.9 illustrates several buffers. If the object is a point, then the buffer becomes a circle of the given radius. If the object is a line or a polygon, then the feature becomes a polygon extended outward in all directions by the buffer distance. The result of a buffer operation in the vector data model is always to create a polygon feature. Buffers are often used to determine if other objects are "within" a certain distance. If they fall within the buffer, then they are closer than the distance. A set of buffers with increasing buffer distances can be used to create a set of concentric circles around a point that may be used to show, for example, zones of travel time (e.g., under 5 minutes, 5–10 minutes, etc.) or noise levels (e.g., over 100 decibels, 75–100 db, etc.).

Overlay analysis was popularized by Ian McHarg in his 1969 book *Design with Nature*,[7] though others arguably should receive some credit for its conception. Although McHarg's implementation used transparent acetate overlays of maps of suitability classes for various uses (e.g., a set of

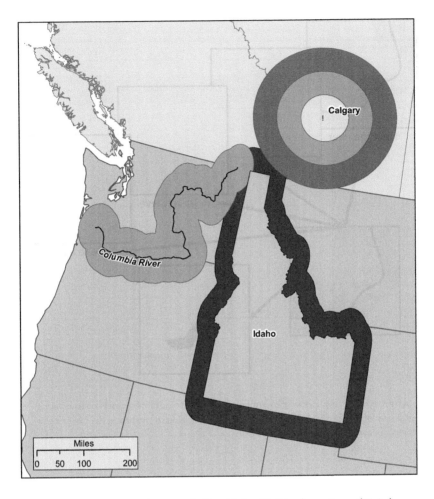

FIGURE 3.9. Fifty mile buffers around a line (Columbia River), a polygon (Idaho) and a point (Calgary). From this process it is possible to tell that the most northwesterly point of Idaho is less than 50 miles from the Columbia River and that the northern border of the state is more than 150 miles from Calgary.

criteria that determines the suitable locations for a particular land use), implementing overlay analysis in the GIS is an extremely powerful and multifaceted tool. The idea is to overlay a set of maps (or GIS layers), each showing a different attribute, that are of the same place and geographically aligned so that coincidences in space can be identified. Overlay can be done with both raster and vector data, though again their implementa-

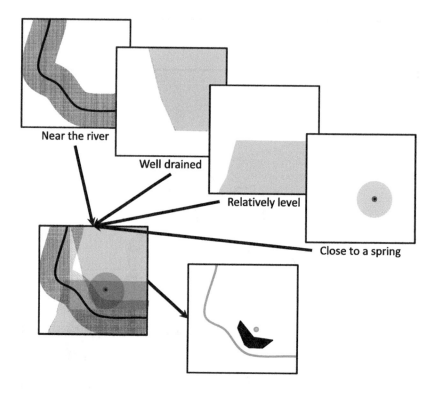

FIGURE 3.10. A simple example using overlay of a site suitability analysis to locate a good site for establishing a settlement. The black region in the box in the lower right shows the area that satisfies all the criteria.

tions are very different. With the proper mathematical functions, overlay can be used as a means of implementing areal interpolation. It can also be used to make holes in a dataset by cutting out the data for areas that fall within an overlaid set of polygons. You might do this to cut out the lakes in a map of population density to show that in those places there are no people. It can also be used to merge two datasets together so that the original polygons are broken up into many smaller polygons, each with attributes from all overlaid layers. Figure 3.10 illustrates a simple suitability analysis using overlay.

Zonal functions are one group of a collection of functions used for analyzing raster data. This form of analysis is often called map algebra since the objective is to combine a set of raster layers mathematically.

Beginning with a set of rasters, all with the same cell size, origin, and orientation (in other words, covering the same location), it is quite simple to determine a large range of characteristics about places using simple cell-by-cell calculations. Raster analysis generally is far simpler to implement than vector analysis. Many map algebra functions are a form of overlay analysis, but it also can be carried out on a single raster by comparing the values of adjacent cells. Say, for example, you have two rasters of population counts. To calculate population change, you simply subtract the values in each cell in one raster from the corresponding cell in the other. Map algebra in its original formulation[8] categorized functions into:

- Local—calculations on individual cells, such as difference, sum, and mean;
- Focal—calculations on cells in relation to their neighbors. For example, using a raster of elevation values, calculate the direction of flow by comparing the elevations of adjacent cells—water will flow to the lower cell;
- Zonal—calculations such as area and total count when one raster in the overlay stack identifies zones as contiguous areas of cells with the same value; and
- Global—which produce a single value summarizing the raster.

All good GIS textbooks provide descriptions of the most important basic spatial analysis techniques. Grouping them all into a small set of categories provides a way to appreciate the full spectrum on offer without taking a GIS course. Although there are many ways to do this grouping, David Unwin and David O'Sullivan, authors of *Geographic Information Analysis*,[9] a required text on the bookself of any informed GIS user, suggest one such categorization:

- Spatial data manipulation is the core functionality provided by GIS. Techniques are diverse and include area and distance measures, buffer, point-in-polygon, overlay, and raster to vector conversion. These are the basic GIS skills.
- Spatial data analysis includes those techniques that are descriptive and exploratory such as point pattern analysis, viewshed analysis, and determining the shortest paths along a network. Advanced analytical techniques might include spatial data mining, exploratory spatial data analysis (ESDA) and spatialization (though many of these are statistically based so they might be placed in the following category).

- Spatial statistical analysis produces measures that help users determine whether results are unusual or unexpected, thus statistically significant. These are generally more advanced techniques that require an understanding of basic statistics and an awareness of the limitations of traditional statistical techniques in the context of spatial autocorrelation. (Traditional statistical techniques assume that each sample is independent, whereas Tobler's First Law reminds us that almost all geographic data are spatially dependent.) Spatial statistical techniques include those such as spatial regression (including its related indexes such as Moran I and Geary C), geostatistics (a set of techniques that includes kriging) and trend surface analysis.
- Spatial modeling produces predictive results. Models can be used, for example, to predict movement of goods, people or water, or to predict changes in the landscape. Predictions can be used in historical contexts to compare one's theory as to change over time with what is recorded. Neural networks and agent-based modeling are among these advanced modeling techniques.

Besides the seemingly endless number of spatial analysis techniques, there are many different ways to combine and implement them. The magic of spatial analysis is in figuring out how to apply the generic techniques to your specific problem. Since many of the techniques come out of the environmental sciences, when we apply them in the spatial humanities, the result can be surprisingly innovative and thought-provoking. For example, the focal map algebra method to calculate flow of water described above can also be used when the raster is cost of transportation. This technique might be used to investigate the failure of the antebellum South to build roads, at least before the investment in railroads, because of the ready availability of (low cost) navigable streams and rivers, or the movement of migrants along a line of least resistance, as when southern colonials moved west via the Cumberland Gap. Therefore, when learning to use GIS, some focused effort should be expended on exploring the range of techniques available so that a number of possible methods for investigating a specific problem can be considered.

Often a set of techniques will be chained together into a processing model so that the output from the analysis of input layers flows into subsequent analytical steps. Figure 3.11 depicts one such processing model. Such models make it easy to visualize the analysis stages and most com-

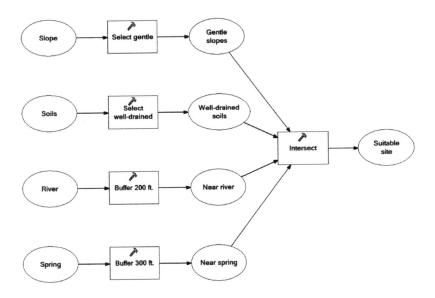

FIGURE 3.11. A multi-step model for chaining together the series of spatial analysis functions performed to create Figure 3.10.

mercial GIS now provide functionality to store processing models, along with the specification of input data, for reuse, fine-tuning, and sharing.

DEALING WITH ERROR AND UNCERTAINTY

One final topic must be mentioned, albeit briefly. All representations of the world, whether they are maps, attribute tables, or GIS data layers, are incomplete and contain errors. Error arises in every step of our work from conceptualizing our world, to representing it, measuring attributes, storing values and analyzing data. Uncertainty is a result of this error, though it can also arise from, to name a few, lack of semantic clarity, imprecise measurements, missing data, temporal inconsistencies (using data collected over several different time periods), or the use of proxy measures for data we cannot collect (such as using level of education, which is available from the census, as a proxy for sophistication of voters). In some circumstances it is possible to estimate the amount of error that is likely in data; GPS measurements are often reported with +/− meters or feet estimates and most statistical analyses produce error estimates.

However, in many circumstances it is not possible to identify all the sources and impacts of the unknown error. When working with GIS in the humanities it is essential always to ask questions about the quality of the data, the fidelity of the representation used, the inherent error, and whether the resulting maps or spatial analysis truly represent the world that has been modeled. Many of the issues mentioned in this chapter highlight sources of error and uncertainty in the use of GIS. The best we can do is be aware that error exists, understand the quality and "fitness for use" of the data we are using, attempt to reduce the level of error and uncertainty in our work as much as possible and/or determine the effect of the probable error through sensitivity analysis, and be honest in how we represent the reliability of our results.

CONCLUSION

GIS and spatial analysis has huge potential for use in the humanities. While time has traditionally been the primary dimension of focus in most humanities disciplines, the spatial dimension has always been lurking in the data collected. In addition to the widely accessible mapping tools now available on the Web (such as Google Maps, Google Earth and Microsoft Virtual Earth), the advanced tools provided by modern GIS have finally reached a level of maturity and relative ease of use that make them appropriate for the non-expert to begin exploring those hidden spatial data. As the other chapters in this volume demonstrate, there is an unlimited frontier awaiting those willing to venture forth into this new dimension.

NOTES

1. Karen K. Kemp, "Introduction," in Karen K. Kemp, ed., *Encyclopedia of Geographic Information Science* (Thousand Oaks, Calif.: Sage Publications, 2008).

2. H. Couclelis, "People Manipulate Objects (but Cultivate Fields): Beyond the Raster-Vector Debate in GIS," in A.U. Frank, I. Campari, and U. Formentini, eds., *Theories and Methods of SpatioTemporal Reasoning in Geographic Space* (Berlin/Heidelberg: Springer-Verlag, 1992).

3. See, for example, Werner Kuhn, "Geospatial Semantics: Why, of What, and How?," *Journal on Data Semantics*, 3 (2005), 1–24.

4. Ruth Mostern, "Historical Gazetteers: An Experiential Perspective, with Examples from Chinese History," *Historical Methods* 41 (2008), 39–46.

5. Waldo Tobler, "A Computer Movie Simulating Urban Growth in the Detroit Region," *Economic Geography* 46 (1970), 234–420.

6. Michael F. Goodchild and Nina S. N. Lam, "Areal Interpolation—A Variant of the Traditional Spatial Problem," *GeoProcessing* 1 (1980), 297–312.

7. Ian L. McHarg, *Design with Nature* (Garden City, N.Y.: Natural History Press, 1969).

8. C. Dana Tomlin, *Geographic Information Systems and Cartographic Modeling* (Englewood Cliffs, N.J.: Prentice Hall, 1990).

9. David O'Sullivan and David Unwin, *Geographic Information Analysis* (Hoboken, N.J.: Wiley, 2002).

FOUR

Exploiting Time and Space: A Challenge for GIS in the Digital Humanities

IAN GREGORY

INTRODUCTION

Most information is explicitly or implicitly concerned with theme, time, and space. Much humanities scholarship is concerned with a theme and how it varied over time and/or space. Handling space and time together is difficult, and this complexity has often led scholars to focus on either change over time, the domain of historians, or variations over space, primarily studied by geographers. Arguing that history is the study of time and geography the study of space overstates the divide between the two; however, it is fair to say that history is concerned with the study of periods in the past, while geography is the study of places on the Earth's surface.[1] As time is clearly a central concept to period and space is important to place, the difference between space and time appears to be a central reason for the disciplinary divide between history and geography. History has what Monica Wachowicz calls a time-dominated view, while geography has a space-dominated view.[2] Many academics within and beyond these two disciplines have long argued that the division between time-dominated and space-dominated approaches is counter-productive and a more integrated approach is needed. This has been taken furthest in physics where Einstein's work has moved the discipline beyond seeing space and time as separate but related concepts to a single concept of space-time. Arguments also have been made by geographers, historians, and others that to understand the human environment we need to move towards a concept of space-time.[3]

Geographic Information Systems (GIS) originated from a space-dominated approach and has long been criticized for its poor handling of time.[4] In theory, a GIS layer represents how a theme associated with the Earth's surface varies over space. This variation is either continuous, in the case of raster data, or a series of discrete vector features such as points, lines, or polygons. There is no explicit concept of time within a GIS layer. It can be handled either as an attribute associated with each feature or by having separate layers for each different time, both which provide workable but limited ways of handling discrete time.

GIS has been less frequently criticized for its poor handling of space but here again it has limitations. There are two basic ways in which a GIS represents space: through Euclidean geometry that allows it to calculate straight line distances and angles on a uniform surface, and by topology which allows it to understand which lines connect to which other lines and which polygons lie on either side of a boundary.[5] At its core therefore a GIS has only a crude quantitative representation of space, but this nevertheless allows the GIS to create sophisticated models of places on the Earth's surface and to understand the relationships between them.

In addition to time and space, the third component of data is theme or attribute. Here again, GIS is limited. It originated in the Earth sciences, a quantitative data-rich environment in stark contrast to the humanities which tend to be restricted to a small number of limited and complex sources that are normally textual rather than numeric.

Thus a critic could claim that to a GIS, geography is simply coordinates and connectivity, information is simply quantitative data, and time is very poorly represented. There may be some justification in this yet it can equally, and perhaps not contradictorily, be argued that, although crude, the GIS data model has the potential to breathe new insights into our understanding of space and time in the humanities, in particular by allowing us to understand how change happens in different ways in different places. Increasingly applied research is emerging to support these claims. This essay will review what space and time are in terms of their impact on GIS, how they are related to each other, how they are important to history and geography, and why our failure to manage both in tandem has limited the technology's potential in the humanities. The paper will then move on to explore how GIS can help us to gain an understanding of space and time, as well as the limits that exist to that understanding.

TIME, SPACE AND SPACE-TIME: THEORETICAL UNDERPINNINGS

Although superficially simple, time is a problematic concept. At the heart of conceptualizing time is a paradox identified by Augustine, who stated that he understood what time is as long as no one asked him, but if asked what it is, then he did not know.[6]

Six ways of conceptualizing time can be identified: linear, calendar, cyclical, container, branching, and multiple perspectives.[7] Of these, linear time is the simplest. Time is represented as a unidirectional continuum that flows in a straight line from infinity in the past to infinity in the future. Calendar time is sub-divided into precise times and dates allowing distances in time to be measured into the future or into the past. Cyclical time stresses that time is not always linear but flows in cycles. The most obvious manifestation of this is the changing seasons; business cycles provide another example. In cyclical time it is acknowledged that one winter will be similar to the previous one but will be considerably different to the summer that lies between them, even though summer is closer to winter in linear time. Container time builds on this idea by splitting time into discrete periods of varying resolutions and granularities such as days, months, and years. Some containers, such as hours of the day or months of the year, can occur cyclically, while other containers, such as years, eras, or epochs, are linear. These containers can have some surprising effects, for example, December 31st may occur in the same week as January 1st but is in a different year.

In the above examples time follows a single trajectory but this need not be the case. The fifth way of conceptualizing time, branching time, occurs where multiple lines evolve from, or lead to, a situation at a single point in time. It is usually represented as a tree structure similar to an evolutionary tree or a family tree. An evolutionary tree starts with a single species at one time and evolves through many branches over time. A family tree can do the opposite, in that multiple ancestors led to an individual at a single point in time. Branching time is often used to forecast scenarios that may occur in the future as time evolves according to different possibilities. Finally, time can have multiple perspectives. This happens when the time an event occurred in the real world is not necessarily the time when it is recorded as having happened in a database or on a paper source. In computer science terminology, the time when an event actually occurs is called real time, event time, or valid time. The time it is recorded as happening is called transaction time, system time, or database time.[8]

There are some clear similarities between these representations of time and the way we represent space, although some important differences also exist. As with time, space can be linear in that it can flow from point A to point B. There are two ways in which it differs from time: first, we can move through space in any direction rather than just from the past to the future, so space flows in two or three dimensions rather than just one. Second, while time flows from an infinite past to an infinite future, space on the Earth's surface is fundamentally finite. In the same way as the calendar is used to split time into discrete units, so space is subdivided into arbitrary units such as meters and miles. Cyclical time means that periods some distance apart temporally may be assumed to have similar characteristics. Spatial concepts such as urban and rural, upland and lowland, and desert and forest follow a similar structure. Just as time is containerized into arbitrary units that may or may not meaningfully represent these cycles, so space is also containerized. These containers may attempt to follow actual characteristics of space, such as urban areas or a particular soil, vegetation, or climatic zones, but they may also be more arbitrary such as parishes, districts, or counties. The last two representations of time, branching and multiple perspectives, are perhaps less applicable to space as they rely on the fact that time, unlike space, always flows from the past to the future.

These ideas of space and time follow the Newtonian idea that time is a dimension that is separate from but similar to space. More recent work in physics, based on the ideas of Einstein, Minkowski, and others, has led to the understanding that it is over-simplistic to consider space and time as separate entities. Instead, physicists conceive of time as a fourth dimension that is inextricably linked to space. Rather than space and time being two separated concepts, it is possible to conceive of space-time as a single combined concept in which a four-dimensional geometry can be constructed. In two-dimensional space, Pythagoras' Theorem is used to calculate the distance between two points. The distance is the square root of the sum of the squares of the distance in x and y. This can be extended to three dimensional space by adding height, z, and four dimensional space-time by adding time, t. It is important to note however that even in the equations of Einsteinian physics time does not function in exactly the same way as a fourth dimension of space would be expected to. Thus even in this conception, where space and time are closely linked, time cannot be seen as the same as space. As will be seen below, this difference becomes even more important in the human experience of time and space.

One way of demonstrating the significance of the concept of space-time is by the use of Minkowski's light cones. The speed of light is critically important to the physicists' conception of space-time because nothing can travel faster than light. Because no impact from an event can travel faster than the speed of light, there is a four-dimensional cone that spreads through space-time from the point at which the event occurred that potentially could have been influenced by it. In places in space-time beyond this, it is impossible for the event to have had any effect because the event is unknown and unknowable. The places in space-time where the event can have had an impact are called its future light cone. Similarly, a past light cone can also be drawn that encloses places in space-time that could possibly have influenced the event.[9]

The concept of space-time has become well established in physics. The obvious next question is: how do concepts devised to explain how particles behave when they approach the speed of light affect the humanities? John Kelmelis attempts to apply the physicist's view of space-time to the world as we experience it by replacing the speed of light with what he terms causal propagation.[10] Kelmelis argues that the spatial extent of a process is a function of the magnitude of the process, its temporal extent, the time that has elapsed since the process started, the velocity at which it can propagate, and what he terms an attenuation factor. While this is an interesting idea, it relies on the mathematical approach of the physicist which is not the type of approach that will find favor in the humanities. Instead we need to move towards a more abstract understanding of how space and time interrelate and why this is relevant to the humanities.

TIME AND SPACE IN HISTORY, GEOGRAPHY, AND THE HUMANITIES

As previously identified, history is the study of times or periods in the past. Many historians are concerned with change over time and often tell the story of how events unfolded during, for example, a war or a monarch's reign. Not all histories are about change over time, however, for the subject matter may be concerned with a theme associated with a particular period such as land-use in Medieval England or governance in Ming Dynasty China. Similarly, geography is centrally concerned with place and how places interrelate over space.[11] In the same way that historians need not be concerned with change over time, geographers need not be concerned with variations over space. Indeed there are a range of ways in which geographers can conceive of space. At one extreme, space is the medium that enables places to interact and limits the way in which they can do so. In this conception, space is thus concerned with transport and its associated costs in distance, time, or money. At the opposite extreme, space allows multiple places to co-exist even though there is no interest in how they interrelate. A study that compares an English village with a French one requires space since without it there can be only one place. While this may seem obvious, it actually gets us to a major reason as to why space is important, namely, space allows difference and diversity.[12] The way geographers study variation across space has some similarities to the way that historians study change through time, but the differences are also significant. A historian might study a continuous linear chronology in which one event leads onto the next, while the geographer is more likely to study discrete containers in space that may or may not interact. Thus, while there are similarities in the way that space and time are handled in these two disciplines, there also are differences.

History/time and geography/space are explicitly linked in two distinct fields of study: historical geography and time geography. Historical geography is concerned with a place or places in the past. The concept of space and time used may be as simple as studying a single place in a single period of the past or, at the other extreme, may be about how a process operated over space and time in the past. Space and time may thus be explicitly studied or merely provide the background to the study.[13] Time geography is more explicitly concerned with movement, flows, and diffusion across space and time.[14] The classic study in time geography is T. Hagerstrand's study of the diffusion of innovation over space and time;[15] other studies have focused on scales as diverse as the movement of individuals during a typical day[16] or the spread of epidemics over continents and decades.[17]

The humanities encompass many more disciplines than history and geography. They also include literary studies, religious studies, classics, archaeology, and many others. History and geography are, however, important to most of them. A work of literature might be set in a particular time and place such as Wordsworth's Lake District or Dickensian London. Travel writing tends to concentrate on movement through space, while biographies tell the story of an individual's journey through time. Epic tales such as Tolstoy's *War and Peace* tell of journeys over space and time. Religious studies and the classics have strong historical and geographical influences being set in, for example, the Holy Land in biblical times, Ancient Egypt, or Roman Britain. Archaeology is perhaps somewhat more complex in that it brings together precise measurements of space and time when conducting digs,[18] with much broader concepts of history and geography when interpreting archaeological sites.

TIME, SPACE, AND SPACE-TIME IN THE HUMANITIES

William Cronon highlights an example of where two approaches, one that has time without space and one that has space but not time, are applied to the same subject area.[19] Cronon was interested in the growth of Chicago in the American West and explored two theories of how land-use develops from wilderness, through increasingly dense settlement, to the city. The first approach, where time is examined without space, was developed by historian Frederick Jackson Turner.[20] He saw the American West as starting with "savagery" inhabited only by Indians and hunters, then traders arrive, then cattle ranching, then extensive agriculture based on crops such as corn and wheat, then more intensive agriculture, and finally industry. At each stage the settlement pattern becomes denser leading to cities in the industrial age. At the same time a different theory was developed by Johann von Thunen, an economist. Von Thunen saw land use as being determined by the cost of transporting goods to market. Thus, there would be a ring of intensive agriculture such as dairy cattle and orchards around the city. As distance from the city increases so agriculture becomes more extensive until the final ring of cattle ranching. Beyond this would be wilderness as it would no longer be viable to transport goods to market. This theory is concerned with how continuous spatial change, in the form of rising transport costs, leads to discrete spatial zones in the form of concentric rings of different land uses. Remarkably, both theories come up with the same pattern: Turner has us moving from the wilderness, through increasingly intensive agriculture, to the city as we move through time from the past to the future. Von Thunen has us following the same pattern as we move across space from wilderness to the city. Turner has time but no space, Von Thunen has space but no time.

The key point of this is that without both space and time we are unable to gain a proper understanding of the phenomenon under study. Turner is unable to account for the fact that some areas become cities while others remain agricultural or even wilderness. Von Thunen is unable to explain how the city developed chronologically and the impact this had on changing land use. Although similar, the theories also have a clear contradiction. According to Turner, intensive agriculture is a forerunner to urbanization. According to von Thunen, intensive agriculture can only exist if there is an urban area large enough to support it. Therefore for a fuller understanding of historical and geographic processes we need to understand both space and time.

This practical criticism identified by Cronon is similar to the more theoretical argument developed by Doreen Massey, who explored the role of space and time in geography.[21] She argues that too often geographers either look at "timeless geometrical models based on two dimensional planes" or study short-term processes.[22] She argues instead for an approach that looks at historical development as potentially open-ended. Time is needed to tell the story of how an individual place developed to become what it is now, however without space there is only one story and thus the risk that this is seen as the only possible story and the inevitable story. Space is needed because it allows for more stories and thus for a diversity of experience. Thus, Massey argues, space-time in geography is concerned with exploring multiple trajectories through space and time to allow complex stories of how places change to be told.

The case for a view of the world that incorporates space and time together has also been argued in historical geography. John Langton used systems theory to identify two approaches to the study of change: synchronic analysis and diachronic analysis.[23] In synchronic analysis a system is allowed to reach equilibrium under a set of parameters. The system then undergoes a change in parameters. Once the system reaches a new equilibrium state in all its parameters the two sets of parameters can be compared to explore change over time. Diachronic analysis "attempts to trace the origins of particular elements of the system and the interrelations and then follows the evolution of the way they function, cutting across a successive series of synchronic pictures of the system."[24] Langton goes on to argue that diachronic analyses are necessary for explanation since systems can rarely be assumed to be in equilibrium. In short, to understand time it is not sufficient simply to compare isolated snapshots, but rather to follow processes through a continuous sequence of change.

To draw the work of Cronon, Massey, and Langton together, in an ideal world we would use spatial and temporal data to explore how a phenomenon has evolved over time, not by comparing two snapshots but by looking at continuous change. In doing so the aim is not to identify the story of how the process evolved but to use different places to explore the different ways in which the phenomenon could occur differently.

SPACE AND TIME IN HISTORICAL GIS

From the above discussion it is clear that a strong case can be made as to why it is important to have an integrated understanding of space and time. To date, however, complexity has undermined attempts to do this. Gail Langran argues that when confronted by information that contains space, time, and theme, the response has traditionally been to measure one, while controlling the second, and fixing the third.[25] The decennial population census, for example, measures the distribution of the population by controlling space, which is divided into discrete arbitrary units, and fixing time to a single day or period. A topographic map also fixes time to a survey date, theme is controlled by dividing the features to be included into relatively simple classes such as vegetation types and contour intervals, and space is measured by defining where these features lie. The question is: can GIS, with its quantitative focus, its Euclidean and topological concepts of space, and its lack of temporal functionality allow us to move to a more holistic representation where all three components are not only measured, they also are studied in an integrated way?

Historical GIS is perhaps the best developed part of GIS in the humanities.[26] One of the early manifestations of historical GIS was the development of national historical GISs, databases that link a country's census data to the administrative boundaries for which the data were collected. Because national historical GISs usually cover the nineteenth and twentieth centuries in an integrated database these systems provide a representation of all three components in the original source: theme in the form of statistics, space in the form of administrative units, and time in that they cover many decades.[27]

Although these databases are expensive and time consuming to produce, they do not by themselves produce a contribution to knowledge in the humanities. An example of the kinds of analysis that these databases enable is provided by my work on infant mortality in England and Wales from the 1850s, when comprehensive data first become available, to the 1900s.[28] Infant mortality is the number of deaths among babies before their first birthday, usually expressed per thousand births. Until the nineteenth century, infant mortality rates were typically around 150 per 1,000 births. The late nineteenth century saw the beginnings of a decline that lasted throughout the twentieth century to fall to rates of less than 5 per 1,000 today. The conventional explanation for this decline is that infant mortality was primarily an urban problem exacerbated by rapid urbanization and industrialization in the nineteenth century which led to appalling conditions in overcrowded and unsanitary towns and cities. Rates started to fall as a result of the public health movement which led to improvements in sanitation, provision of fresh cows' milk in cities, and efforts by midwives and others to improve parenting skills especially by mothers. This orthodoxy originated with the Victorians themselves and is still largely accepted today.[29] There is thus only one story of change over time, the urban story which is seen as the one that led to the national story. Attempts to critique this interpretation have suffered because of the inability to handle space and time effectively. For example, Clive Lee used county-level data,[30] but this scale is too coarse to provide urban/rural contrasts. Naomi Williams and Chris Galley used a small number of sample districts,[31] but it is difficult to tell the extent to which these are representative of broader trends.

Having data from the Great Britain Historical GIS available meant that I was able to use the maximum amount of spatial and temporal detail available from the original sources, the Registrar Generals' Decennial Supplements. The source data were available aggregated spatially into 635 registration districts and temporally into six decades for the period from 1851 to 1911. These data were then re-aggregated in two different ways to try to identify the major trends within them. The first approach involved aggregating the districts into eight classes based on their population density to identify the trends according to different levels of urbanization. Infant mortality rates over the six decades for each class were then plotted. This approach revealed that infant mortality in rural areas was already falling by the 1860s, while urban areas largely followed the national trend. This finding is important because if infant mortality decline was caused by urban solutions, such as improved sanitation and midwives, then why did rural areas decline earlier and more steeply?

The second form of aggregation explored whether there was a split between core and periphery in which rates in the core, London and the South-East, should be low and then rise with distance. To explore this theory, I divided the country into buffers 25km wide centered on London and calculated the infant mortality rate for each buffer based on the births and infant deaths of the districts whose centroids lay in each buffer. The results showed that over six decades the country appeared to become more polarized between core and periphery, with rates in the core outside of London showing the biggest improvements and rates in the periphery lagging well behind. This pattern had not previously been recognized.

Combining these two patterns shows that rural areas generally had lower rates than urban ones in the 1850s. Over time the largest improvements were found in the rural south-east while areas in the rural periphery showed some of the lowest improvements or, in some cases, actually got worse. Urban areas overall improved at around the national rate. Establishing this pattern relied on imposing aggregations on the data into a rural/urban hierarchy and a core/periphery hierarchy.

An alternate approach to aggregation is "to let the data speak for themselves." This was done by using areal interpolation to standardize data from every decade on a single set of registration districts to allow direct comparisons over time.[32] Once this had been done, it was possible to explore how many infant deaths were averted by the change in infant mortality between the 1850s and the 1900s. Two highly contrasting patterns were found. The total numbers of deaths averted reflected a pattern that was largely urban suggesting that where cities improved a large reduction in the numbers of deaths occurred and, where they did not, large numbers of infants continued to die. If the number of deaths averted is mapped as a proportion of the deaths that would have been expected had infant mortality rates remained at their 1850s level, then a very different pattern emerges. In this instance, the biggest improvements were found in the rural south-east, while the rural periphery of the country showed the lowest improvements. This discovery confirmed patterns found through aggregation.

This analysis shows that as well as the urban story of infant mortality, there were at least two others: the rural south-east story of low rates that started to decline early on and showed the highest overall declines, and the rural periphery story of low rates in the 1850s that largely failed to improve or became worse. Here we encounter a major limitation of historical GIS work of this type. Using a database of disaggregated information for every birth and infant death over a span of sixty years, the GIS is able to do an effective job on identifying what happened, where it happened, and when it happened. The GIS is, however, unable to say why this happened. One thing that is clear is that the existing explanation, that infant mortality decline started as a result of urban improvements is, at best, simplistic and, at worst, wrong. Rural mortality decline started before the urban improvements were implemented and led to the biggest rates of decline. While mounting an effective challenge to urban solutions causing infant mortality rates to decline, the GIS is unable to provide its own explanation. The traditional GIS approach to doing this would be to perform a statistical analysis on a large range of other quantitative data associated with agriculture, employment, social conditions, and so on. These data are simply not available in sufficient detail in a consistent form over time to allow this to be done effectively.

How then can we move towards explanation? It would seem that traditional historical approaches are more likely to be of use than GIS approaches. Detailed work in local record offices to see how local areas' economies, societies, agriculture, and health care changed over the sixty years seems to be the most likely way to explain the changes in infant mortality in local areas. The problem is that this type of work remains highly labor intensive and is thus only viable for a small number of local areas. Here the GIS can complement this approach because it is able to assist in locating similar areas and, by contrast, dissimilar areas and in this way guide further research. In this way the broad-brush GIS approaches complement the detailed local case studies using traditional history methods.

BEYOND QUANTITATIVE HISTORY

While the approach outlined above shows much promise, it is limited to quantitative data. Most historians do not use quantitative sources and, as previously established, the humanities encompasses disciplines other than history and these other areas are often even less numerate. To make spatio-temporal GIS approaches more generally relevant, we must be able to take large volumes of text, convert these into a GIS based on points, and then analyze them through space and time. To date this has not happened but some progress has been made in this direction.

The first task in turning a text into a GIS is to extract the place names. From a linguistics perspective this is relatively straight forward as place names are proper nouns and algorithms to identify these are well developed. This provides a list of proper nouns that may be place names. These are then filtered to remove words that are obviously not place names, for example proper nouns that are preceded by "Mr." or "Ms." can be considered as people's names and are thus not considered.

The second stage is to attempt to georeference the remaining proper nouns so that coordinates can be added to them. This again is technically straight forward using a gazetteer, effectively a database table that provides a coordinate for every place name that they contain. Where the spelling of a proper noun taken from the text is identical to the spelling in the gazetteer and there is a one-to-one match the noun is initially assumed to be a place name. Where there is no match the most likely explanation is that these are not place names, however, these need checking as a significant number will be variant spellings. Where these are found they need to be matched to the gazetteer versions of the place name which, in turn, enables us to use these techniques to enhance gazetteers. Common place names such as Bangor, Newcastle, and Newton are problematic as they require some user input to disambiguate the actual place referred to from the others in the gazetteer. A final problem that can occur is that proper nouns coded as place names, such as "Victoria," are found to refer to people rather than places. Where this happens we need to return to the first stage and improve the filters on proper nouns to remove these cases. For example, cases where the words such as "Queen" precede "Victoria" this is clearly a reference to a person rather than a place. This means that georeferencing a text tends to be an iterative process whereby the linguistics learns from errors located once the proper nouns have been linked to a gazetteer. Early experiments conducted on an 800,000 word corpus[33] of newspapers published in London in 1654–55 have suggested that the approach can be effective in georeferencing large bodies of text relatively quickly and accurately.

Once coordinates have been identified for place names, then the act of converting them into a GIS is a trivial task. While this gives a map of the text, it does not give any indication of time or of what is being said. For some sources, such as newspapers or diaries, adding time is relatively straight forward. In others it is more complicated and may not be possible. Adding a theme is an easier problem as linguistics algorithms allow us to see what words occur near to a place name and these can be classed into categories such as those relating to war, famines, the economy, and so on. As long as it can be assumed that if a word of this sort occurs near a place name, the source is discussing this theme in relation to this place, then it is possible to map what the text is saying about a place. While this assumption is slightly simplistic in a large body of text, it is unlikely to be overly problematic.

Research into the georeferencing of corpuses is still in its early stages but clearly has the potential to allow large volumes of text to be structured by space, time, and theme. Georeferencing texts opens up the possibility of analyzing these data through space and time in a way that has to-date been the preserve of polygon-based quantitative data. The task, once again, is to make best use of the thematic, spatial and temporal information to challenge whether the orthodox story that has been told from a source is the only one, or indeed the most important one. Again, this approach should complement, not replace, the more in-depth approaches that have been traditional in the humanities.

CONCLUSIONS

GIS allows us to locate large volumes of data in space and time and this has the potential to offer significant new understanding across the humanities. To date, most of the progress on this has taken place in quantitative history, but as textual sources become increasingly important in GIS we can expect similar progress in the non-quantitative humanities. In a spatio-temporal analysis it is important to move away from a space-dominated or time-dominated view to a more integrated approach. Whether this can be described as space-time is a moot point as there are major differences in the way that we experience time and space, as well as the way each is represented in the sources used in the humanities. Nevertheless, Massey's approach of using space to identify multiple stories of development over time seems a productive one, and one that is set to challenge many long standing assumptions. Geoff Cunfer's challenge to the hypothesis that the Dust Bowl was caused by over-intensive agriculture is a second example of this.[34]

There are limits to what we can expect GIS to achieve on its own. As a technology it is a superb tool for simplifying and describing large volumes of data, however its explanatory power, as with any technological solution, is limited. This result means that GIS approaches need to be used in parallel with the more traditional forms of interpretation, with one informing the other. For example, the GIS identifies where the major stories are happening, case studies try to explain the stories, and the GIS can then say where the different explanations seem to be valid and where they are not.

Further developments still need to be made to make this process simpler and more effective. Visualization and analysis techniques are still very much either space dominated or time dominated. In visualization we typically have the map or the time-series graph, but integrating these is difficult. One approach is through the use of animations, but with choropleth data these are very limited in what they can achieve. The static choropleth map already relies on simplifying continuous attribute data to a small number of classes in order to make the spatial pattern clear. Adding the temporal is not effective without further simplification of either the attribute or the spatial, so making animations effective communication tools is probably only possible with relatively simple datasets. With dot-density maps, this may be less problematic. Density smoothing techniques can be used to make point patterns more understandable and adding temporal density to spatial density is an approach that has already been implemented in subjects such as crime mapping. In this way we can create relatively understandable animations of how point patterns change over time. Here again we have a problem. Although these techniques cope with space and time effectively, theme is a further issue and including it is difficult. Some humanities-based examples of animations have been created,[35] but the degree to which they offer new insights into the topics under study has yet to be established.

Improved exploratory data analysis techniques may offer some solutions as they are able to provide alternative solutions as to how to summarize complex spatial and temporal patterns. Again, there is the problem that analysis techniques tend to be space dominated or time dominated. There are plentiful spatial analysis techniques and time-series analysis is even more established, yet there are few if any reliable spatio-temporal analysis techniques. Uni-variate techniques that identify spatio-temporal clusters are one example, but what is desirable are multi-variate approaches that can identify how different variables related to each other over space and time. It is unclear, however, how these would work and how much use they would be in the humanities.

GIS provides a data model that assists researchers in the humanities to understand space and time simultaneously, but this does not provide a complete solution to the problem. The challenge remains for the humanities researcher to use the framework provided by GIS to make best use of the spatial and temporal aspects of the research question under study and in this way to develop solutions that are appropriate for the data and paradigm that they are using.

ACKNOWLEDGEMENTS

This paper benefited from support from the British Academy under grant SG46005 "Literary mapping of the Lakes: A pilot for a humanities GIS." The work on georeferencing texts is being conducted in collaboration with Dr. Andrew Hardie, Department of Linguistics and English Language, Lancaster University.

NOTES

1. Alan R. H. Baker, *Geography and History: Bridging the Divide* (Cambridge: Cambridge University Press, 2003).

2. Monica Wachowicz, *Object-Oriented Design for Temporal GIS* (London: Taylor & Francis, 1999).

3. Doreen Massey, "Space-Time, 'Science' and the Relationship between Physical Geography and Human Geography" *Transactions of the Institute of British Geographers: New Series*, 24 (1999), 261–76; Doreen Massey, *For Space* (London: Sage, 2005).

4. Gail Langran, *Time in Geographic Information Systems* (London: Taylor & Francis, 1992); D. Peuquet, "It's about Time: A Conceptual Framework for the Representation of Temporal Dynamics in Geographic Information Systems" *Annals of the Association of American Geographers*, 84 (1994), 441–61.

5. Digital Terrain Models (DTM) and various forms of raster analysis do allow space to be modeled as three dimensional, but these are specialized applications for specific fields of investigation.

6. W. H. Newton-Smith, "Space, Time and Space-Time: A Philosopher's View," in Raymond Flood and Michael Lockwood, eds., *The Nature of Time* (Oxford: Basil Blackwell, 1986), 22–35.

7. See: Christopher Yearsley and Michael Worboys, "A Deductive Model of Planar Spatio-Temporal Objects," in Peter Fisher, ed., *Innovations in GIS 2* (London: Taylor and Francis, 1995), 43–51; Michael Worboys, "A Generic Model for Spatio-Bitemporal Geographic Information," in Max J. Egenhofer and Reginald G. Golledge, eds., *Spatial and Temporal Reasoning in Geographic Information Systems* (Oxford: Oxford University Press 1998), 25–39; and Suzana Dragicevic and Danielle Marceau, "A Fuzzy Set Approach for Modeling Change through Time" *International Journal of Geographical Information Science*, 14 (2000), 225–45.

8. See Yearsley and Worboys, "A Deductive Model of Planar Spatio-Temporal Objects"; Worboys, "A Generic Model for Spatio-Bitemporal Geographic Information."

9. Stephen Hawking, *A Brief History of Time: From the Big Bang to Black Holes* (New York: Bantam, 1988).

10. John Kelmelis, "Process Dynamics, Temporal Extent, and Causal Propagation as the Basis for Linking Space and Time," in Max J. Egenhofer and Reginald G. Golledge, eds., *Spatial and Temporal Reasoning in Geographic Information Systems* (Oxford: Oxford University Press, 1998), 94–103.

11. R. Johnston, *Philosophy and Human Geography: An Introduction to Contemporary Approaches* (London: Edward Arnold, 1983).

12. Massey, *For Space*.

13. See: A. Baker, *Geography and History*; Robin A. Butlin, *Historical Geography: Through the Gates of Space and Time* (London: Edward Arnold, 1993); and Robert A. Dodgshon, *Society in Time and Space: A Geographical Perspective on Change* (Cambridge: Cambridge University Press, 1998).

14. See Don Parkes and Nigel Thrift, *Times, Spaces and Places: A Chronogeographic Perspective* (Chichester, UK: Sage, 1980); Allan Pred, "The Choreography of Existence: Comments on Hagerstand's Time-Geography and Its Usefulness," *Economic Geography*, 53 (1977), 207–21.

15. Torsten Hagerstrand, "What about People in Regional Science?" *Papers of the Regional Science Association*, 24 (1970), 7–21.

16. Joseph Weber and Mei-Po Kwan, "Bringing Time Back in: A Study on the Influence of Travel Time Variations and Facility Opening Hours on Individual Accessibility," *Professional Geographer*, 54 (2002), 226–40.

17. Peter Haggett, "Prediction and Predictability in Geographical Systems," *Transactions of the Institute of British Geographers*, 19 (1994), 6–20.

18. In archaeological digs space and time overlap in another way as time tends to be reflected as depth, a spatial concept.

19. William Cronon, *Nature's Metropolis: Chicago and the Great West* (New York: W.W. Norton, 1991).

20. Frederick Jackson Turner, *The Frontier in American History* (New York: H. Holt, 1920).

21. D. Massey, "Space-Time, 'Science' and the Relationship between Physical Geography and Human Geography"; and Massey, *For Space*.

22. Jonathon Raper and David Livingstone, "Development of a Geomorphological Spatial Model Using Object-Orientated Design," *International Journal of Geographical Information Systems*, 9 (1995), 359–83 quoted in Massey, "Space-Time, 'Science' and the Relationship between Physical Geography and Human Geography," 262.

23. John Langton, "Potentialities and Problems of Adapting a Systems Approach to Change in Human Geography," *Progress in Geography*, 4 (1972), 123–78.

24. Langton, "Systems Approach to Change in Human Geography," 137.

25. Langran, *Time in Geographic Information Systems*.

26. See: Ian Gregory and Paul Ell, *Historical GIS: Technologies, Methodologies and Scholarship* (Cambridge: Cambridge University Press, 2007); Ian Gregory and Richard Healey, "Historical GIS: Structuring, Mapping and Analysing Geographies of the Past" *Progress in Human Geography*, 31 (2007), 638–53; Anne Knowles, ed., *Placing History: How Maps, Spatial Data and GIS are Changing Historical Scholarship* (Redlands, Calif.: ESRI Press, 2008).

27. For descriptions see: Anne Knowles, ed., "Reports on National Historical GIS projects," *Historical Geography*, 33 (2005), pp. 293–314.

28. Ian Gregory, "Different Places, Different Stories: Infant Mortality Decline in England & Wales, 1851–1911," *Annals of the Association of American Geographers*, 98 (2008), 1–21.

29. Robert Woods, "Infant Mortality in Britain: A Survey of Current Knowledge on Historical Trends and Variations," in Allain Bideau, Bertrand Desjardins, and Héctor Pérez Brignoli, eds., *Infant and Child Mortality in the Past* (Oxford: Clarendon, 1997), 74–88; R. Woods, P. Watterson and J. Woodward "The Causes of Rapid Infant Mortality Decline in England and Wales, 1861–1921. Part I" *Population Studies*, 42 (1988) 343–66; and R. Woods, P. Watterson and J. Woodward "The Causes of Rapid Infant Mortality Decline in England and Wales, 1861–1921. Part II" *Population Studies*, 43 (1989), 113–32.

30. Counties in England and Wales are the rough equivalent of states in America. There are approximately fifty of them (exact numbers depend on definition), their size varies enormously, and they usually have both rural and urban areas.

31. Naomi Williams and Chris Galley "Urban-Rural Differentials in Infant Mortality in Victorian England" *Population Studies*, 49 (1995), 401–20.

32. Ian Gregory and Paul Ell, "Breaking the Boundaries: Integrating 200 years of the Census Using GIS" *Journal of the Royal Statistical Society, Series A*, 168 (2005), 419–37.

33. Corpus is the term linguists use for a large body of text.

34. Geoff Cunfer, *On the Great Plains: Agriculture and Environment* (College Station: Texas A&M University Press, 2005)

35. See for example: *TimeMap Project, TimeMap Animations* (in depth): http://www.timemap.net/index.php?option=com_content&task=view&id=124&Itemid=130 (accessed 10 Feb. 2010).

FIVE

Qualitative GIS and Emergent Semantics

JOHN CORRIGAN

The possibilities for qualitative Geographic Information Systems (GIS) rest largely on the prospective successes of humanities researchers in interrogating GIS in a way that will compel its adaptation to humanities data. One way of characterizing what is at issue in that turn as it now has begun is to observe that GIS ontology currently privileges disambiguation in its organization of knowledge, while in the humanities, it is trust placed in the slipperiness of data, in its status as multivalent, equivocal, and protean, that determines the processes of its sorting and analysis. Our imagining an eventual common ground that escapes the dread gravity of this seemingly perennial problem—a problem articulated in many ways since the Enlightenment as a "war" between science and its epistemological competitors—might be well served by our focusing on ways of exploiting opportunities for multimedia GIS. The challenge is to construct a spatial multimedia that is coherent and productive even while remaining emergent. In simpler terms, we need a more fluid and ambiguous GIS.

To suggest that GIS ought to get in touch with its inner ambiguity is to give the appearance of misunderstanding how GIS developed and what role it has played in geographic and information science research (among other areas). GIS has always been a tool of analysis, in the same way that Dr. John Snow's map of the cholera epidemic in London in 1854 was a means to an end, namely, saving lives. Researchers have created the opportunity for more complex and significant analysis by geocoding massive amounts of data and displaying that data in computer generated visualizations. The GIS software that has enabled such research presumes that all data can be named. We know what a church is. We know what a church member is. We can gather statistics about the number of churches of different Christian denominations in a certain place, say, the United States, and the number of people who are church members, and, armed with those numbers, create a GIS of church membership in the United States. We can query the data in various ways to advance understanding about religion and space, and, layering it with other kinds of data, position ourselves to identify linkages between church membership and household income, ethnicity, criminal activity, educational level, and so forth. In other words, we create an exceptionally sturdy platform for analysis, one that can support a broad range of investigation into religion in America. That platform is built out of statistics, which are patterned electronically in response to our queries, according to a grid of latitudinal and longitudinal coordinates. The engine runs on hard data, tagged in such a way as to disclose quantities. This is traditional GIS. It is not fashioned out of respect for ambiguity.

If humanities GIS has a future, it will incorporate at some point early on discussion of what we mean by data and how we name the inventories collected as such. Quantitative GIS coalesced as an artifact of the interplay between an approach to knowledge steeped in the legacies of the social sciences—synchrony, generalization, scientific certainty are three terms that apply in this case—and the mode of data coding (mathematical) demanded by an electronic information system designed to translate numbers into images (although the earliest GIS did not actually generate images). For traditional GIS, the central issue is to get the numbers right, to make certain that the core data is correct. Count the churches carefully. Don't rely on hearsay, out of date surveys, or data suspected of being otherwise corrupted. Quantitative accuracy is crucial to framing a dataset for display within a GIS environment.

Even a triple-checked and verified count of churches will not reliably support analysis, however, unless the hermeneutics that have been employed to construct rationales for datasets are sound. In other words, the collected data can mislead unless it is named in a way that accurately defines it. In our case, are the organizations that we identify as Christian churches all in fact churches? More to the point, has the inventory omitted some organizations that should in fact be included? If we define a Christian church as an association of persons with certain specific theological beliefs and practices, along a fairly traditional line of understanding what those beliefs and practices are, we risk overlooking not only new religious movements but also Christian-tinctured religious organizations that might depart in some ways from standard definitions—from orthodoxy—but nevertheless could in some calculus be included. At the most basic level of rethinking definitions, ought we to include Mormonism as a Christian church? Are Mormons Christian, as most Mormons claim, or are they non-Christian, as many Christian churches claim? In more complex rethinkings of what a church is, should we include organizations such as the Masons or Shriners, which promote religious practice that resembles some Christian rituals and incorporate beliefs that are Christian-flavored if not orthodox? At the most complex level of defining religious associations, is there, for example, an American civil religion that takes its leads from Protestant Christianity and that is practiced in word and deed on national holy days such as the Fourth of July and Memorial Day, or even Thanksgiving? How can the space(s) of such a church be represented in a GIS format that also must be geared to display the location and layered attributes of spatially tidy, more plainly bounded associations such as the century-old Presbyterian church on the town green?

In one sense this is a problem of scale—an issue not unfamiliar to those who invoke GIS in a range of research fields. Are some churches just more churchy than others and can that be reflected in a spatially oriented display? Is the one hundred-member Christian weightlifting club that meets three times a week for prayer and a long evening of pumping-up more or less of a church than the First Baptist congregation of sixty members who meet once a week for two hours of services and coffee? If it shows up in a visual display at all, does it loom larger or smaller, greener or redder, cross-hatched or solid? Can a GIS engineered to disclose scale be adapted to accurately reflect the space and place of the varying kinds of organizations that we recognize as Christian churches? In another sense, this is a problem that cannot be solved by attention to scale alone. The entire process begs the issue of how we define religion, a social unit, religious practice, and tradition itself. It elicits the kind of educated suspicion that humanists bring to their observation of their subject matter—people and culture—and calls for a way of complicating the visual display of data that can preserve some measure of the ambiguity inherent in data about people and culture.

Another aspect of the qualitatively-oriented research typical of the humanities is the deeply embedded discursive orientation to blend assertion of value with the presentation of data. Of course, all branches of research—whether in the sciences, social sciences, or humanities—employ rhetorical strategies in the process of presenting evidence that supports analysis. It is possible to persuade through the design of charts in ways not entirely dissonant from campaigns to persuade that are schemed in prose. Nevertheless, humanities researchers, by virtue of their commitment to language in broaching interpretation, operate differently than colleagues whose vaunted argumentative paradigms rely on quantitative means. Numbers, not language, are more likely to form the core of explanation in fields outside the humanities. If government and foundation funding patterns over the last twenty years are any indication, the regime of numbers seems to carry the day over words. And there are of course workshops like this, where we specifically have gathered to discuss qualitative research in the humanities vis-à-vis quantitative research in other fields that have experimented with GIS. In the interests of sparking our imaginations to revisionings of GIS for the humanities, we ought to consider the differences between these paradigms of interpretation, and especially how researchers draw conclusions when quality is privileged. We ought to take seriously the claim of George Lakoff and Mark Johnson that argument made in prose is usually made in metaphor and that metaphors are weighted with different kinds of meanings—which is why we embrace them. More importantly, metaphor also implicitly invites judgment. Woven into the language of a critical interpretation of Poe's short stories, or an article on ancient Greek philosophy, or an analysis of Hindu sexual practices are judgments about how we are to view each datum that is presented. When an article on Native American healing details the contents of the shaman's medicine bag by stating that, "feathers, insect carapaces, stones, patches of dried skins of animals, and various plant matter are crammed into the pouch," it ought to come as no surprise to us later that Native healing practices are judged slipshod and ineffective. Fortunately, and perhaps paradoxically, the multivalency of language can also allow us to glimpse subjectivities, so that humanists who are playing at the top of their game manage to judge, and press conclusions on readers, at every step of the way at the same time that they relent in nailing down the interpretation for good. In the current (and disturbingly trendy) lingo, humanities scholars claim to "gesture towards" a conclusion rather than state it. They do so because in many instances the case already has been made in the linguistically coded presentation of evidence along the way.

Humanities scholars care as much about nuanced defining of data categories as they do about collecting the data that will be organized under those categories. Humanities scholars also are in the habit of blending the presentation of data with judgments about it—humanities journals do not publish articles explicitly arranged as contiguous sections of overview, literature review, thesis, research method, data, and conclusions. One final observation about how humanities scholars work is useful here: we prefer dynamic renderings of culture rather than static ones. Humanities scholars, with certain exceptions, tend to steer themselves away from analytical models that orient persons to data as if to a snapshot. One "quality" of data that good humanities researchers look for when they gather "qualitative" data is complexity. I already have "gestured towards" this at the beginning of my report, but some detail here might prove salutary. It is still worthwhile to count steeples, as my colleagues in the field of American religious history say, because we need to know how many churches there were in a place to be able to write a religious history of it. But steeple-counting alone is not of much use in understanding the matrix of feelings, forces, and follies that form the backbone of the religious lives of people in a place. Complex data is required for that: broader data about material culture and built environment to begin with, but also the traces of peoples' lives that can be gleaned from letters and diaries, photographs and drawings, wills and epitaphs; topological data—did they live on top of a mountain or by the sea?; records of criminal activities—was there a high murder rate?—and marriages and courtships—who was engaged to whom and who died unmarried?; epidemiological information and data about in-migration and out-migration; political leanings and social status; cycles of economic contraction and expansion. In short, understanding religion in a place means keeping all of these balls in the air at the same time, and making possible a perspective that is dynamic, that changes—like the balls in relation to each other as they move through the air. Complex data is dynamic data in that sense. It is data that is characterized by interaction between its various parts, which massage each other continually in ways that alter our understanding of what we are looking at. When we remember that the central trope of historical writing in the last half century is the irony of history (as opposed to the positivism of the nineteenth century), we can appreciate more of how this concern for complexity has shaped interpretation in the humanities. Humanities scholars are not allergic to drawing conclusions in analyzing data. It may be the case, however, that humanities scholars like to draw a suite of conclusions at the same time, and not all of those conclusions might comport exactly with each other. This fascination with complexity, it seems, ought to draw scholars to multimedia GIS, which is potentially well-situated to disclose complexity on a scale that does not seem available in prose accounts.

QUALITATIVE RESEARCH AND EMERGENT SEMANTICS

But this is to get ahead of the task a little bit. GIS indeed does hold promise for the kind of interpretive projects that humanities scholars undertake. But in order to envisage some possibilities therein it is worthwhile to reflect on how GIS has been employed in the humanities already, and one of those ways has been in historical studies. What historical GIS has managed, though imperfectly, is some integration of the standard synchronic GIS—one that discloses relationships between data in fixed moments of time, such as we might observe in the Great Britain Historical GIS, the China Historical GIS, and the Salem Witch Trials projects—with initiatives that experiment with a diachronic approach, as might be found among some of the various works of the Electronic Cultural Atlas Initiative (the French and Spanish Missions in North America GIS is one example). Such experimentation necessarily involves refinements in software, and a tool such as TimeMap, for example, or ArcView extensions (or Toolbook, MapInfo, etc.) can enable some integration of the temporal axis with the spatial. The display of change over time that is observable in such a GIS, especially when presented as an animation, can disclose much about the way in which people and culture move through space, appear and disappear, and exist in relation to natural environments. It also can turn up—in ways that are not as readily apparent in synchronically oriented GIS—relations (sometimes asynchronous) between elements of the datasets, or layers, that are not seen as such so easily—not seen as dynamically—when viewed as frozen moments, even if the temporal dimension is fairly well developed (i.e., lots of moments in time). With a view to what has been accomplished in historical GIS, and what remains as desiderata, we might think harder about how a temporal axis is more than just a measurement of elapsed time. We should assay conceptualizations of movement through time, as well as space, as the intrinsic and necessarily unstable relationship between attribute data. In other words, we should reflect on how our data is, in a sense, alive, and incapable of disclosing human realities if we force it to stop moving. Data, like butterflies, become something else when they are, as the saying goes, "captured," pinned to a foamcore board. A humanities GIS chases the butterfly rather than nets and pins it.

All of this admittedly is beginning to sound a little mystical—butterflies, living data, contradictory conclusions, definitions that do not define—but in order to appreciate ways in which the qualitative core to humanities data might, in the context of current technology, be realized in a GIS, we have to take a broad, even philosophical, view. In order to glimpse how multimedia can be integrated into a GIS, for example, we have to keep in mind the slipperiness of humanities data, and imagine its role in a GIS as one that both reinforces and challenges the picture that is obtained in a synchronic, quantitatively-rich visualization.

There are different kinds of multimedia that offer possibilities for humanities GIS, in terms both of the range of media that can be incorporated and styles of display. Texts, graphics, video, photos, animation, 3-D renderings, audio, and virtual reality identify the majority of possible components for Multimedia GIS (MMGIS). There are a number of issues relating to the integration of such components in a MMGIS that are worth discussing, but we might begin by thinking about our cognitive capabilities as users—at least as we might understand cognition in a general sense—and how construction of a humanities GIS can be informed by examples of humanities research in non-electronic modes. We should take care in our imagining a humanities GIS that we do not suppose our perceptual and cognitive capabilities as embodied humans will naturally expand to take in and process multimedia data without limits. Just as automobile drivers in an urban environment must select—consciously or unawares—what stimuli matter in reaching the destination (red lights matter, but do the smells of things burning?), so might the user of a MMGIS have to make choices. In other words, there is such a thing as information overload. A MMGIS is not going to prove itself if the spatial multimedia presentation frustrates actual cognitive integration. Spatial coordination of data is one way to organize information, and an effective way. But how much and how presented? And to what extent can multivalent images and sounds, the ambiguity at the core of humanities research, be let off the chain to fully inform the user without confounding him? Is there a way to organize multimedia in connection with the relative precision of statistics and spatial coordinates? How does a movie clip from *Gone with the Wind* make sense of the line of the Chattahoochee River? Do rasters and representations, vectors and visualizations comport? Can we read the text of a Confederate soldier's letter home, listen to a recording of camp singing, digest spatially coded statistics about nineteenth-century food production, process a temporally enhanced dynamic GIS display of changing battle lines, and watch Katie Scarlett O'Hara Hamilton Kennedy Butler rely on the kindness of strangers all at the same time?

Humanities researchers, as I have previously observed, juggle different kinds of data. Writing in the emergent field of the history of emotion—a slippery topic even for a humanities scholar (or especially for a humanities scholar, depending on how one looks at it)—is a case in point. While trying to recover the traces in text of how a group of persons in a certain place at a certain moment in history felt about their children, the researcher must remain aware of a multitude of contextualizing cultural backgrounds, ranging from religious ideas and economic orders to the history of education, gender and sexuality, and epidemiological cycles. Typically, this process involves moving back and forth between sets of texts and notes, keeping some things in mind about gender, for example, as one reads about cholera, or remembering details about a community status order while one reads about religious services—or studies portraits of ministers. The problem is always the same: how to focus and to unfocus at the same time. It is the same problem I have every time my university library informs me that they are recalling a book I have checked out. I say the name of the book a few times and then scan my several shelves of checked-out books looking for that title; but to do so I have to read the titles on the spines of the books I am looking at, and soon enough—after a hundred books or so, sometimes much sooner—I have forgotten the title I am seeking. I lost my focus on the title of the recalled book while I unfocused to read the spines. Humanities research is not as frustrating as looking for a recalled book, but the process of investigation is in some ways similar.

In research, we build a firewall against entropic forays such as are represented by book-hunting by starting with a thesis, or sets of theses, that pull our attention and imagination back into orbit around what matters. We start, in a sense, with a story, and we refine and redirect it as required by the data we discover. Where the humanities differ from the sciences—where researchers also generally start with theses and stories—is that in the humanities, we work with a more fluid semantics. Scientific research, as linguist Roy Harris recently has argued, operates with a reocentric semantics, *The Semantics of Science*.[1] That is, scientists deploy a semantics in which words stand for real things rather than our ideas about things. In the humanities, words about even those things that we seem most certain about—such as ourselves—change as our ideas change. They change as our situatedness changes. So when the scientist Buckminster Fuller announced that "I think I am a verb" he was invoking a humanities semantics that is best understood as an emergent semantics. Such a semantics enables and thrives on a shiftiness in the meanings of things. Rather than reify, it ambiguates. Nevertheless, it is a platform for the production of knowledge, and, like science, is negentropic, not entropic.

To return to the issue at hand: how do humanities scholars find a grounding for their research initiatives in an emergent semantics and how does that clue us to what a humanities MMGIS might look like? Humanities investigators have built and continuously rebuild an emergent semantics out of our experiences of having to focus and unfocus at the same time, of having to juggle datasets or collections of artifacts. Our experience of continuously altering the names of things, or at least to redefining terms, apparently has been positive. Our work is guided and conditioned by our determination to tell a story, and that, hopefully, keeps us from going too far adrift. What all of this adds up to should be no surprise: a fluid semantics is how the humanities are grounded. And a MMGIS/3-D-VQGIS will be of most benefit to the humanities when its architecture incorporates an emergent semantics—through utilization of ontology-based metadata or macro languages or in some other form—as well.

An emergent semantics in the GIS rendering of a scene might be described most simply as a capability to recode data on the fly so that it can be brought into interaction with other data in new ways as the rendering develops. So, at the base technical level, the GIS architecture should be capable of responding to the humanities scholar's need to orient datasets to each other in different ways as analysis proceeds alongside observation. The ideal interactive responsivity in this case is an artificial intelligence. At present the layering of layers of assorted data, the inclusion of a temporal dimension, the visualization of scale, and the embedding of text, audio, and images, among other things, allows for the construction of a qualitatively-rich environment of geocoded data. Plenty of good analysis can come through engagement with such a presentation of information. The trick is to build a dynamic relationship between the data by establishing some measure of responsivity between the spatial coding, temporal referencing, and linguistic tagging of data. In other words, the three key referencing systems—space, time, and language—might be engineered in such a way that changes in one ripple into the others. How would qualitative data be resited in space or time or both when it is retagged? How could unexpected intersections of spatial and temporal references for data lead to retagging that data? How can a GIS do this work of humanities researchers in a way that maximizes the sheer power of electronic information processing and display?

By way of hypothetical example of how a humanities GIS operating with a simple version of an emergent semantics might function, we could imagine how the development of a religious Atlantic World might be dynamically rendered. Let us assume we have robust data on the movement of a wide range of artifacts and people around the Atlantic World (Europe, Africa, the Americas) since 1500: religious ideas, clergy, church records, sophisticated demographics, print culture, trade, the African diaspora, military affairs, disease, education, emotional states, political ideologies, dress styles, polemical literature, and so forth. Alongside this data available as layers and sub-layers—some of which has a qualitative aspect—we have texts, images, audio and other kinds of data that are largely qualitative. With all of the layers deployed, we examine the geocoded slave trade data that includes attributes about slave religious orientations. We query "Africa," "Christian," "indigenous," and "Muslim" and note that during a certain period slaves are brought to the New World from East Africa rather than West Africa, and that the area is strongly oriented towards Islam. If we looked closely at the layers made visible, we might also see that this particular period also happens to be a time when economic expansion is strong, political ideology is shifting to more democratic models, clergy on board ships are decreasing, disease is spreading more rapidly, there is an increase in the despair people feel in certain places, the military is stretched thinner because of conflicts in Europe, and so forth. But we are mere humans who cannot process so many correlations very well, or even notice them in some cases. The job of the GIS is to remember the relationships between all those layers and sublayers so that when we make similar queries of different times and places the words we use are systemically linked with tags for other significant data that emerged, unnoticed by us, in a previous query. The GIS remembers the ways in which certain data, from a range of layers, correlated with the forced migration of Africans who were Muslims. Querying with "Muslim," "Africa," "Slavery" and some other terms consequently displays relationships among data in that part of the Atlantic that the GIS had noticed (and we didn't) in investigating the slave trade. The GIS reminds us, visually, that engagements between Christians and Muslims correlates with the spread of disease, the production of anti-Islamic art, cycles of boom and bust in the silk trade, a spike in the publication of polemical literature, increasing despair, and so forth. GIS displays a visualization of correlations between data that we had not anticipated because through its reading of linguistic cues it mimics our thinking and can in fact think faster about the millions of data in the base. Although we did not query "despair" it shows up in the visualization because there is enough correlation among other parts of the data—which we had not noticed previously—to warrant its inclusion. On a subsequent search, the pattern repeats itself, human query priming the deployment of an emergent semantics which grows richer and more complex with each query.

In the more ambitious version of this, the computer notices correlations that we do not and undertakes the task of retagging other data to reflect correlations it has discovered, and then, in cascading fashion, it experiments with the relation of retagged data to other geocoded data, producing from that operation new sets of relations that require further redefinition, and so forth. This is what humanities scholars do, but no humanities scholar will ever be able to do it as quickly or as thoroughly as a GIS powered by artificial intelligence with access to online gazetteers. Of course, there are plenty of questions to answer in the meantime, not the least of which is how linguistic codes can be created for such an enterprise. But the possibilities, even in the near future, are intriguing.

MULTIMEDIA

The primary reason for the integration of multimedia presentation into GIS, from a humanities perspective, is that a MMGIS represents human subjectivity more efficiently than statistically-grounded visualizations. Quality in this sense is understood to be closer to the subject while an emphasis on quantity objectifies, although that is only a working definition. The role of multimedia in GIS is to soften hard data, to ambiguate it. In some cases, images, for example, can reinforce patterns visible in a more traditional GIS. A set of photos of traffic jammed bumper-to-bumper in an urban center easily enough drives home the meaning of maps coded to represent massive volumes of commuter traffic. Add the sound of horns honking and the marriage of the display of human frustration with geocoded road use statistics is consummated. More important, however, are instances in which qualitative data challenge the seeming realities indicated in the lines and colors of a GIS. The point of a multimedia GIS should be the creation of a representation of culture that illustrates its complexities and contradictions. What if we had location-linked video that showed that traffic moving quickly and smoothly through town? An audio component that registered just an occasional horn? Cars stopping for pedestrians? People who did not seem to be in a mad rush to the office garage? What would that mean? How does that kind of multimedia challenge the analysis (of human behavior as well as traffic flow) that might be drawn from a purely statistically driven GIS?

In thinking more concretely about how multimedia enhances analysis in GIS by fostering a useful ambiguity, we might consider the Salem Witch Trials Web site. The site provides a map of the homes of the accused witches and their accusers during the early 1690s. That map, in its inventorying of persons actively involved or implicated in the witchcraft panic, suggests the possibility that the household locations of those persons form a pattern upon which analysis can rest, and in fact that was the conclusion of historians Boyer and Nissenbaum (who interpreted the witchcraft episode as a case of conflict between parties representing two different parts of the town).[2] But when the points representing the households of accusers and accused each are linked to text of transcribed trial testimony, the differences between the situations of the various accusers becomes much more apparent, and what might have appeared to be a telling pattern of spatial distribution of witches and complainants is made ambiguous. The apparent patterning in space of the persons involved is challenged—a closer look at each reveals that the tags "witch" and "accuser" that were used to construct a map are more complex and tenuous than originally thought, for example—and explanation of the witch hunting campaign less sturdily grounded in analysis of the locations of participant households. In observing this, we can foresee how the integration of media into a GIS would serve the larger role of challenging, on a separate but related front, the tendency of statistically grounded GIS to overlook ambiguity. As such, it would play a central role in the construction of an emergent semantics.

CONCLUSION

A humanities GIS will advance as researchers are able to fashion a GIS with an emergent semantics. The qualitative bias to the humanities requires that ambiguity be accounted for in the organization of information about people and culture. An emergent semantics, which is only truly possible if linguistic leeway is built into GIS ontology, will enable utilization of the power of the machine to detect patterns in spatially-coded data and, in acts of primitive awareness, intervene in human-initiated exploration of spatial data to enhance analysis in ways ordinarily too complex for human cognitive capabilities. At the same time, such patterning must be made interactive with multimedia that can ambiguate statistical data and challenge its authority in the process of analysis.

NOTES

1. Roy Harris, *The Semantics of Science* (New York: Continuum, 2005).
2. Paul Boyer and Stephen Nissenbaum, *Salem Possessed: The Social Origins of Witchcraft* (Cambridge, Mass.: Harvard University Press, 1974).

SIX

Representations of Space and Place in the Humanities

GARY LOCK

INTRODUCTION—FORMS OF REPRESENTATION

Space and place are complex and elusive concepts[1] so any representation of them (which according to my dictionary is something that "corresponds to, or is in some way equivalent or analogous to") is immediately contentious. Or is it? Through this chapter I would like to intertwine notions of landscape with those of representation, partly because landscapes are obviously spatial at one level but also because landscape is a humanizing theme which draws together many of our disciplines.[2] Considerations of landscape have also brought to the fore a tension which is central to forms of spatial representation: both landscape and maps are at the same time a representation of material phenomena and an interpretation formulated around and within subjective meaning. So, to answer my own question, it is possible to have representations that are not so slippery and contentious, indeed we all regularly use maps, or these days more likely sat-navs in our cars, which perform in an acceptable functional way via an accepted code of symbolic representation that attempts to avoid subjective interpretations (although we still get lost!). The oft used quote from the influential Chorley Report,[3] "GIS [Geographic Information Systems] are the biggest step forward in the handling of spatial data since the invention of the map" is interesting here as it implies a technology that can go beyond the traditional map into new realms of representation. Twenty years on though, the challenge is still whether we can take subsequent arguments and theoretical constructs and percolate them down into practice, particularly the difficult practice that involves GIS.

There is obviously a historical thread here and my own discipline of archaeology has a long history of close connections and shared traditions with geography plus commonalities with the rest of the humanities. Not least is the interpretative trajectory from early empiricism through a quantitative revolution and the quest for objectivity, ultimately dissolving into a crisis of representation and the uncertainties of postmodernism. Archaeology tentatively boarded the GIS bandwagon in the late 1980s, and with the publication of *Interpreting Space*[4] experienced a massive uptake of the technology over a relatively short period on both sides of the Atlantic. As with geography, though, it hasn't been an entirely smooth ride as by the early 1990s archaeology had already largely rejected the quantification paradigm and the adoption of GIS was seen by some as a retrograde step back toward reductionism and environmental determinism. Early applications such as statistically based predictive modeling of site location and territorial analysis of resources were accused of being detached and pseudo-objective when compared to contemporary writings such as: "Landscapes are created by people—through their experience and engagement with the world around them. They may be close-grained, worked upon, lived in places, or they may be distant and half-fantasised."[5]

As with other humanities disciplines, GIS is now fundamental to much archaeological practice such as purely representational cartography, more analytical cartography based on thematic mapping, and various analytical approaches both vector-based and raster- or cell-based. Within this range, however, there is variation within the objective/interpretative tension mentioned above, perhaps here more usefully brought into focus when re-framed as observation/inhabitation. In the rest of this paper these two extremes, and the continuum between them, form the basis for the discussion of representation: on the one hand is cartography and mapping, already well embedded within GIS practice, and on the other the lived-in world of phenomenology and "non-representational" theory, unlikely to ever be satisfactorily addressed through traditional 2-D GIS, although virtual worlds hold some hope. Other issues and theoretical currents also need to be taken into account: scale, for example, is central to representation and to understanding data and interpretation; time and historical depth are embedded within landscapes but notoriously difficult to represent in GIS. The idea of "deep maps" attempt to go beyond traditional phenomenology, while theories of "practice" and the idea of "taskscapes" all offer ideas and approaches that are both a challenge and an opportunity for theory-aware GIS practitioners in the humanities.

FROM OBSERVATION TO INHABITATION

SPACE-BASED REPRESENTATIONS

Maps and atlases have provided the spatial backbone for historical, archaeological, and geographical research since the disciplines first developed. Using the spatial primitives of point, line, and polygon, GIS and other spatial software are now commonplace in providing simplified Cartesian representations of what is considered important, and therefore worth mapping, in the physical and cultural landscape. Mapping is embedded within observation and empiricism, an early paradigm reflected in the importance of fieldwork, for example, Carl Sauer's belief that "geography is first of all knowledge gained by observation."[6] The belief that landscape was composed of objective, material facts and things that could be measured and represented was typical through the first half of the twentieth century. An emphasis on fieldwork and observation was also central to the approach of W. G. Hoskins,[7] the father of English local history, although his concern was with landscape as history rather than as geography. Although both of these scholars went far beyond mapping by developing their own schools of interpretation, Sauer's cultural landscapes and Hoskin's historical particularism, the objectivity of the drawn spatial record was given primacy.

That maps are not objective representations but rather the product of a specific culturally positioned point of view and set of power relationships is now well established.[8] A traditional two-dimensional map, whether hand drawn or digital, is not a good representation of inhabited place. Both Sauer and Hoskins provided rich textual descriptive support for their maps and, as discussed below, it is incorporating this extra "sense of place" that is one of the challenges for GIS. Another challenge is whether or not Cartesian space can be adapted to incorporate a more humanistic sense and understanding of distance, direction, and position. While coordinate systems provide the quantitative basis for GIS, our own biological and cultural sat-navs work on qualitative relationships such as "next to," "in front of," "behind," and "a little way past the supermarket."[9] This more topological approach to representing space is illustrated by children's maps of their routes to school and also by the illuminating insight into the complexities of map making provided by the Parish Mapping Project of the environmental group Common Ground.[10] The importance of these maps in understanding place is based on the notion of "local distinctiveness" and of what is important to people who live there and encounter aspects in their daily lives that are important to them. It is the fluidity and the understanding of qualitative spatial relationships that is the challenge here to developing GIS methodologies role in humanities research, as illustrated by the single example of the *Barrington Atlas of the Greek and Roman World*.[11] This massive accomplishment of 200 scholars took over 12 years to complete and has resulted in 102 beautifully drawn color topographic maps representing the Classical world from ca. 1,000 BCE to AD 650 together with detailed information for each map on CD-ROM. While we should not underestimate the aesthetics and pleasure inherent with owning and using paper maps and atlases, and appreciating the skill that underlies their production, this atlas does raise questions fundamental to the development of GIS within humanities research, not least those concerning scale and historical/temporal depth.

Although scale is a complex issue, it is something often taken for granted and not well theorized or discussed in humanities research, for example, in archaeology.[12] We can recognize two different scales: "analytical scale" is the purview of research, the recognition and analysis of patterns, while "conceptual or phenomenological scale" is the world of human interactions in daily life.[13] The tension created here through GIS usage is that the technology is good for regional scale representation and has, therefore, tended to be used at that scale of mapping and analysis. The small urban polygons in *The Barrington Atlas* represent towns and settlements and lack the multi-scalarity that, in theory, GIS offers because within those urban areas are buildings, rooms, open spaces, the material world of the everyday. This is important because the micro-scale dynamics of daily living contribute to, and create, macro-scale phenomena such as social reproduction and cultural change which we map and analyze as humanists. The material world and the social world are merged through the inheritance of social norms, culture, and tradition that are both multi-scalar. At the same time this occurs in a recursive relationship so that people are created through their material and social worlds while at the same time creating those worlds. Here there is a constant transformation between social structure (society) and the individual although it is the material world that enables us as individuals and social groups to ensure persistence at temporal scales longer than is humanly possible.[14]

This raises the question of whether social change as reflected through the material world, i.e., change through time, can be represented within GIS and the answer has to be one of doubt. The data layer structure of GIS, and temporal coding within attribute data, enforce a categorical structure onto time which is essentially a continuous phenomena, a limitation of the technology identified a long time ago although little progress has been made since.[15] Of course, this limitation is application dependent, and there are many instances where the representational limitations of scale and time are acceptable, and indeed forcibly argued. For example, Historic Landscape Characterisation (HLC) is a systematic GIS-based mapping of the historic elements within the English landscape on a county by county basis.[16] The polygon net coverage records various aspects of historical information but within a single layer rather than using specific time layers:

> just as landscape is a seamless whole geographically, so too it is seamless chronologically. The whole of the past resides in the present, seen or unseen, known or awaiting discovery. (HLC) ... condenses time into a single layer, which is, in the strict sense, the only landscape we can ever have, the product of our perception today, in the here-and-now.[17]

Relevant to thinking about historical depth and temporal representation within GIS is the work of Werner Kuhn, a specialist in geoinformatics, who suggests that we do not put enough emphasis on process and that we lack semantic models and taxonomies to even start thinking about them properly. He differentiates between *endurants,* things that ARE in time, and *perdurants,* things that HAPPEN in time, the former participating in the latter. A way forward for humanities GIS may be to link these concepts with those of the material world and social practices and the recursive relationship between the two as mentioned above.[18]

There is no doubt that vector data does allow a range of analytical techniques that should be of interest to the spatial humanities. Although buffering can be used with other data structures, it is popular with vector data, for example, the exploration of distances from roads and towns as part of the study of attitudes to the Civil War in 1850s Pennsylvania and Virginia.[19] The layer structure of GIS also allows the deconstruction of the physical world into elements that can be re-classified and re-configured through layer-based analysis. For instance, the complex categories within soil maps can be re-classified and combined with other information such as proximity to water and altitude to produce a new layer representing agricultural potential. This approach is moving toward representing a humanistic understanding of landscape based, in this instance, on farming practices and what will grow where rather than formal soil classifications. Another area of vector-based analysis which holds potential for the humanities is network analysis, introduced in pre-GIS days by physical geographers[20] and now used extensively by utilities companies and for GIS-modeling of the viability of new supermarket locations for instance. The interest in network analysis for the humanities is based on the possibility of social relationships being represented and modeled through networks of connections.[21] Symonds and Ling show how tenth-century eastern English trading and communication networks can be modeled based on the distributions of known towns, quantities of pottery, rivers, and roads.[22] More interestingly, perhaps, is that they go beyond the network to consider how early medieval concepts of time and travel are different to quantified distance measurements, a representation of space as a network of interrelations.

Probably the next move along the representational scale from observation to inhabitation is centered around attribute data and the GIS functionality of databases, querying, and data management attached to geo-objects, that is again a particular strength of a vector data structure. This traditional strength of the technology underlies many applications in the humanities including inventories such as national and local monument records, and census databases. A varied suite of analyses accompany these approaches allowing a range of visualizations of attribute data through choropleth maps, thematic maps, and cartograms.[23] The warnings given regarding the subjectivity of these displays based on the underlying analytical decisions such as class intervals should be heeded by all humanities researchers.[24]

Returning to the idea of endurants and perdurants and how these can be linked to the material world and social practices, the challenge here is how this can be furthered through GIS functionality incorporating attribute data. This reintroduces the importance of scale because attribute data can be recorded and connected to spatial objects that represent different scales of social reality. Daly and Lock have tried to demonstrate this with their study of prehistoric hillforts in central England where the deposition of pottery sherds within pits suggested a series of practices that could be mapped so that certain combinations of sherd types occur within certain pits within certain sites thus creating multi-scalar connections across a landscape.[25] The practice of breaking and disposing of pottery manifests itself at different scales of interpretation thus linking individuals with social groups of different sizes, and it is the structuring of attribute data that enables this to be displayed through multi-scalar representation.

Traditionally, raster or cell-based data have been seen to offer different possibilities for interpretation and representation to the vector-based approaches described above. The incorporation of raster maps as background, whether modern or historic, is now well understood and the considerable potential of the latter is demonstrated by Rumsey and Williams.[26] They show how the representational value of a historic map can be greatly increased when integrated with GIS functionality including the spatial registering with modern maps, pseudo-3-D visualization by draping over a Digital Elevation Model (DEM), and overlaying other information including attribute data. (DEM is a digital representation of ground topography or terrain; it also is known as a Digital Terrain Model, or DTM.) Perhaps the greatest potential for moving along the representational axis toward inhabited landscapes, however, is the utilization of cell-based analysis to model visibility and movement. Both can be point-to-point in the form of line-of-sight and least-cost paths or areal when viewsheds and cost-surfaces are involved. This functionality has been available within GIS since its introduction, and after nearly two decades of applications it is time for their impact and future in humanities-based analysis to be reviewed. There is a strong whiff of technological determinism that surrounds these techniques, for both are relatively easy to perform without any understanding or discussion of the underlying assumptions, technical or theoretical.

The initial, and lasting, attraction of visibility and movement studies is that they are an attempt to avoid the so-called "God-trick" of viewing landscapes from above in a detached and pseudo-objective way thus reinforcing the object/subject divide. While these techniques do not claim phenomenological significance, although they have been accredited with incorporating perception,[27] at least they do attempt to position an individual within the landscape and to, therefore, humanize otherwise dead digital models. For representation through visibility and movement models based on DEMs to become more fully embedded within humanities GIS, the technical and theoretical assumptions underlying them need to be more fully explored and explicated.

The characteristics of the DEM used are fundamental to any analysis, with resolution being critical. A 50 meter DEM, for example, which is not unusual at the landscape scale, will impose serious limitations for movement and visibility. Developing technologies are offering much improved resolution and airborne LiDAR (<u>Li</u>ght <u>D</u>etection <u>a</u>nd <u>R</u>anging), for example, could revolutionize the representation of landscape detail with resolutions of ca. 1.5 meters,[28] although coverage is patchy at the moment in many countries and the computing power needed to process the raw data is considerable. Whether DEMs are interpolated from point or contour data or acquired already processed, it has been recognized for a long time, although rarely considered to be of importance, that different interpolation algorithms will produce very different final models.[29] It is questionable whether this matters anyway when what we are trying to model are qualitative and imprecise characteristics. As humanists, it is not clear to what extent we need to be explicit about the quantitative technology underlying our qualitative thinking and, ultimately, producing the models that we are thinking about.

Visibility studies have increased in sophistication from simple binary viewsheds, through cumulative viewsheds to banded and directional viewsheds. These are attempts at modeling the complexity of vision and visual perception and a series of issues and problems have been recognized which can be categorized as pragmatic, procedural, and theoretical,[30] together with ongoing technical assessments of the accuracy of such techniques.[31] Undoubtedly the more nuanced representations of visibility offered through fuzzy, banded, Higuchi, and directional viewsheds have much to offer over the simplistic binary in or out model, although any application has to be incorporated within a wider theoretical understanding of landscape and its meaning.

Just as a visibility model is ultimately dependent on the resolution of the DEM it is based on, so too are movement models but with interesting differences. It is generally argued, as implied above, that a DEM should be of the finest resolution possible as both visibility and movement solutions are calculated and assigned on a cell by cell basis to see or represent "reality" more closely. This assumption is questionable for movement as people don't usually wander aimlessly around a landscape based on small-scale topographic decisions but move with intentionality toward a known destination. This directional bias can be built into least-cost-path algorithms, a form of anisotropic approach, although not solely based on cost, but usually including slope and force. It may be worth exploring whether average slope/cost values assigned to larger cell sizes represent intentional movement more closely as the walker's gaze and concentration are fixed on the distant destination rather than on the detailed characteristics of the terrain within the next meter or so.

A final raster-based technique that again introduces a new approach to representation is that of agent-based modeling. This form of simulation is not widely used in the humanities but its potential for introducing a modeling and understanding of human behavior is shown by Lake.[32] This work simulates the decision making of hunter-gatherer groups based on knowledge of various resources available at different places within the landscape at different times of the year. The new elements here are a representation of behavior and the incorporated process of learning, modifying behavior and individual and group agency.

TOWARD PLACE-BASED REPRESENTATIONS

In a suggested hierarchy of geographic visualization and representation,[33] the lowest level is 2-D mapping, then comes the 2.5-D perspective view of the DEM, including animated sequences and other additions to display, where the user has the limited control offered by viewpoint determination. At the top of the hierarchy is Virtual Reality (VR) and VRGIS which sees a merging of VR and GIS with a range of user experiences based on the level of immersion, from viewing on a computer screen, through enabling devices like head mounted displays and fully immersive CAVEs. These technologies have inevitably been linked with the theoretical understandings of place and have been claimed as a move away from observational representation toward a representation of inhabitation, a dissolving of the subject/object dichotomy. Central to this has been an interest in phenomenology, attempting to apply the ideas of Heidegger and Merleau-Ponty and the notion that we all experience the material world through our embodiment, "being-in-the-world" and dwelling, that the human body is a given that in some way can link the present with the past. Whilst phenomenology is widely recognized as having been important in the move away from purely representational understandings of landscape in cultural geography and other humanist disciplines, VR attempts to operationalize it have fallen short of being a complete solution.

The problems with phenomenological claims for VR have been fully explored,[34] and have been seen to perpetuate the subject/object split between the person and the landscape as it is still based on detached gaze. It has been claimed that the "bones" of a landscape, the general topography, material elements and their spatial relationships, are constants and moving around and between them creates an experience that is similar now as in the past and even though we can never reach culturally embedded meaning, VR provides a starting point for discussion, thinking and dialogue.[35] This may be so but the relationship between body and the material world is much more complex and reflexive; "perception" and "cognition" are not everything:

> Despite the timely advent of the body in social studies, one often gets the feeling that the human body is the only flesh of the world and that this spiritual lived-in body continues to roam around rather unconstrainedly in an intentional world held together almost solely by human cognition.[36]

Nigel Thrift has suggested that phenomenology is but a part of "non-representational theory" and that theories of practice and agency based on the early work of Giddens and structuration theory[37] and Bourdieu's notion of *habitus*[38] provide an understanding of how people create their material world but are at the same time created by it. Participating in embedded cultural practices, even, and especially those that are routine and create the structure of daily life, are according to Thrift, the "absorbed coping" and "engaged agency" that dissolve the subject/object divide and make us and the material world one.[39] Another element of Thrift's argument, and one that is particularly relevant to claims for VR, concerns the construction of knowledge which is not a passive process but one of practice, "knowledge-as-practice" is what provides meaning for the material world, we need to engage with it rather than gaze at it.

There are, of course, many different strands to the "inhabitation as practice" approach to landscape. And if collectively they indicate a move away from representation, this offers a major challenge to humanities GIS and visualization technologies. Ingold[40] has developed the notion of "taskscape" rather than landscape to give primacy to practice in the creation of cultural landscapes; an approach which reactivates the idea that "affordances,"[41] or what the material world enables us to do, is central. An attempt to apply this through GIS is Trifkovic, who uses DEM-based analysis to represent food and other resources and exploitation practices in prehistoric Iron Gates Gorge, Serbia.[42] This is a novel approach which shows the potential of GIS while also emphasizing the importance of scale through using personal biographies to comment on how individuals and groups can construct their own different interactions with landscape.

Non-representational theory is obviously a challenge to the normative practices of humanities GIS and VR, which is increased by the more recent developments of "more-than-representational theory"[43] and the idea of the "deep map."[44] This moving beyond representational theory attempts to work with a range of experiences and information which is perhaps best summarized in the following lengthy quote:

> The focus falls on how life takes shape and gains expression in shared experiences, everyday routines, fleeting encounters, embodied movements, pre-cognitive triggers, practical skills, affective intensities, enduring urges, unexceptional interactions and sensuous dispositions. Attention to these kinds of expression, it is contended, offers an escape from the established academic habit of striving to uncover meanings and values that apparently await our discovery, interpretation, judgement and ultimate representation.[45]

According to Lorimer, landscape, or perhaps more appropriately being human as the creator of landscape, is in a constant flux and state of becoming and not something that can be objectively represented in a detached way as in VR models. An intriguing example of this approach is Pearson's 'In comes I' in which he uses performance "as an alternative medium of representation to cartography."[46] At the three different scales of village, neighborhood and region, he enacts a sense of place through the materiality of the surrounds, historical depth, oral histories, archaeology, geology and a whole range of other forms of understanding so that his resulting "deep map" is a reworked version of thick description.

Moves toward this richness of experience and environment are beginning within VR through the integration of audio and video clips, images and attribute data, as well as perhaps smell and sound, plus "collision points" which enable access to these qualitative impressions as movement around the scene takes place.[47] Serious gaming software and virtual worlds such as AlphaWorld[48] and Second Life[49] offer even more exciting possibilities for the humanities through their being peopled with avatars and the potential richness of their metaverse. These technologies introduce the social since there is no longer a lone inhabitant but one that can build social relationships with others, have a social presence based on a virtual identity, join groups, and get involved in projects that may create new environments in the virtual world. This reintroduces the material (although within a non-material world) through the use of objects as extensions of the body and bodily functions; understanding the world through practices often depends on the material things needed to carry out those practices. It also links with ideas concerning deep maps as qualitative information can be linked to places and as a member of a group you, or at least your avatar, can participate in the construction of place. There are as yet few academic case studies of such commercial virtual worlds, although their potential for teaching and learning is being investigated.[50] Despite this, the phenomenal popularity of these activities has created representations that cause us to question the very nature of reality itself and the interesting conundrum of how can we move toward understanding virtual reality if we don't understand reality to start with.[51]

THE PRODUCTION OF SOCIAL REPRESENTATION
—WEB GIS AND MASHUPS

There is no doubt that making GIS applications available on the Internet, together with rapidly improving Web functionality, adds value and a new dimension to many of the points raised above. There is perhaps a semantic issue here concerning what we actually mean by "GIS" which is being forced by the rapid development of desktop and Web-based mapping software. In the early days of the technology, to be considered GIS software had to have the four subsystems: data input; data storage and retrieval; data manipulation and analysis; and data reporting, with the ability to carry out topological analysis and create secondary data layers.[52] Without the initial zeal of early adopters the boundaries between GIS and other spatial software seem to be increasingly blurred, and this result can only be more so as maps and spatial data become increasingly within the public domain and less the reserve of academics and various professional interest groups. It is no longer clear whether it matters that we adhere to a strict definition of GIS or whether we are happy to embrace "geospatial" and "geovisualization" as second generational understandings of the same issues. These points are not only brought into sharp focus by new and rapidly developing Web applications but also by the shift in developer communities to include mass collaboration.

Historic maps can now be presented in exciting ways that enable a range of analysis as well as increased accessibility to fragile and rare resources. For example, the *Book of Curiosities,* is a remarkable work in Arabic dating to the late twelfth or early thirteenth centuries (although it is probably a copy of an eleventh century original) providing a treatise and series of heavenly charts and terrestrial maps relating to cosmology.[53] Web functionality allows mouse-over interpretations of the Arabic as well as sophisticated searching of the original text, charts, and maps. These different views of the world are important in developing the humanists of the future and this Web site provides Teachers' Packs to show how history, geography, cosmology, and science can be explored through the view of the eleventh century Islamic world.

Much traditional off-line GIS functionality is now available online including layering, georeferenced symbols with attached attribute data, and increasingly complex levels of searching. A good example of the potential, albeit not strictly humanities, is the MESH WebGIS[54] which presents European seabed habitats in detail. Here a whole series of layers including seabed habitat maps, coastlines and administrative areas, physical data (e.g. bathymetry, seabed geology), biological sample data, and images of the seabed can be searched, zoomed and downloaded together with a metadata catalog of seabed studies that ties this work into a wider academic arena. Another important aspect of this project is that it shows how international collaboration can be jointly published in a form that allows constant updating as the research progresses while retaining maximum accessibility. The theme of Web-enabled international collaboration is also central to the work of the Electronic Cultural Atlas Initiative and, within that work, the development of *TimeMap:*[55] a methodology, community, and software that enables interactive mapping of change through time and space and its visualization via animations.

While all of these examples show the potential offered by Web-based applications they are all, to a large extent, developed by specialists and offered as fully formed products that represent a pre-determined view of some aspect of humanities research. There is however, a revolution underway based on an emerging philosophy of the Internet, and certainly one that underpins Web 2.0, that of mass collaboration. There has always been an anarchic element to the Web, but recent and rapidly expanding developments have seen that desire for individuality take on a new social persona and create new forms of socially mediated representation. This has been described as "structured, mass collaborative creativity and intelligence," or the We-think culture ("we think, therefore we are" rather than "I think, therefore I am"),[56] as Charles Leadbeater has labeled it, and which, he claims, represents a massive cultural shift. People are no longer content with the industrial economy of mass production and passive consumption but are creating virtual social and cultural networks based on involvement, collaboration, and creativity. This is enabled by the agency incorporated into new Web-tools so that we can all participate in creating Web resources rather than just being the end users. Leadbeater illustrates a range of areas where the We-think culture is taking hold, including peer-to-peer education, information, entertainment, technical, and commercial, although another characteristic of this new cultural form is that the boundaries between these traditional categories are rapidly dissolving, merging and re-forming. Here, however, we will concentrate on how this new collaborative, social order is beginning to, and is likely to, influence forms of spatial representation.

The philosophy of We-think and associated new developments in collaborative spatial representations are typified by aspects of the new Web 2.0 services and the Geospatial Web.[57] Central to this are mashups, the creation of a new digital tool from the combination of data from more than one source which take the older idea of Web portals to a new level. As many forms of data and much human activity and interest have a spatial component, the availability of Web resources such as Google Earth (and Google Maps) have produced many map-based mashups. There is a whole community of collaboration formed around this, for example see the Google Maps blog[58] which provides tools and advice as well as links to an extensive list of Google-based mashups. As an example, the UNESCO list of World Heritage Sites has recently been added as a Google layer so that the location of all approximately 800 sites is displayed on either a map, satellite image, or composite of both. Map-based mashups, as other types such as video and photographic, search and shopping, and news, consist of three parts. First, the content provider supplies the data through an Application Programming Interface (API) and Web protocols, in this case the location of sites, where it is linked to an immediate photograph and brief description as well as links to the UNESCO database for detailed information. The mashup site is Google, which provides the spatial context, then the data and maps are put together, or mashed-up, and presented through the third element, the client browser.

If presented as a finished article, this would still be an interesting and useful resource in its own right, but being Web 2.0 it enables more interaction through the ability of anyone being able to add text and images. The "official" representation of UNESCO can be altered and enhanced through the additions of the user community to produce a very different representation of World Heritage Sites. UNESCO has a "shared responsibility" with the general public for these sites, which drives its decision to share the data albeit with caution and caveats, as shown by a single, but typical, example in Peru.[59] Peru has ten World Heritage Sites, one of which is Chavin, since 1985, an extremely important pre-Columbian complex of archaeological remains dating to between 1500 and 300 BC. When viewed on Google Earth, its remote location in the Andes is well illustrated but greatly confused by the cloud of user-added photographs around the site, many of them showing incorrect locations, some many kilometers out of position. The issue, of course, is whether this matters or not.

To many it may not, but surely this is taking the uncertainty of postmodernism too far as the location of Chavin is not fuzzy but definite. We are not talking here about spheres of influence or forms of spatial symbolism that may have wide ranging and variable spatial manifestations but, rather, the actual physical remains. This has important implications.

Firstly it dilutes the authority of UNESCO and its data and introduces unhelpful uncertainty. Secondly it challenges notions of accuracy and precision that have been central to academic disciplines such as archaeology, geography, history, and others that work with landscapes and spatial data. While I fully accept that alternative understandings of space, distance, and location can be valid in different circumstances, for example, the Parish Mapping Project mentioned above, I do not think that applies here as in other formal inventories that record position.

It may be that with time this will resolve itself as these virtual communities are self-regulating and self-validating. The Panoramio photographs include "comment" and "misplaced" facilities, so in theory it is possible to influence and even alter their locations. Contributed photographs and comments add to the shared experience of Chavin and help construct the social web through this collaborative production of information. Allied to this, and relevant to representations of space and place, is the process of tagging, an old term and idea that has been revived as social tagging within Web 2.0 applications. A tag is a keyword or term attached to a piece of information, whether a picture, location on a map, block of text, video clip, or whatever, thus enabling classifications and searching. Tagging raises issues of semantics and thesauri that traditionally have been imposed on users. Underlying the evolution of Web 2.0, tagging is the problem of understanding geospatial language (together with the acceptance that we do not understand it) and the idea that acceptable terms will emerge as being "popular" through the practice of users. This has become known as "folksonomy" (a social taxonomy, social tagging, or collaborative tagging), a bottom-up evolution of language where terms are accepted through their use. Folksonomies are community generated through tools provided at individual site level rather than being part of the underlying WWW protocol and operate through a form of feedback loop.[60]

It is very early days yet for this We-think culture and its possible effects on representations of space and place, although implications are already emerging, not least the tensions between official/fixed/traditional forms of information and socially mediated views of the world. We wait to see whether the photographs of Chavin will ever become linked to its actual location, or whether vernacular descriptions and place-names will become more popular than official versions. It may be that these questions are meaningless anyway if the normative view of the world becomes a socially mediated representation within Second Life.

CONCLUSIONS—REMAINING REPRESENTATIONAL

In this chapter I have used landscape as a metaphor for the social and cultural complexity of being human—both in the way it can be used to represent the past but also, and perhaps more forcibly, the present where any representation is located. In his recent account of the changing understandings of landscape, John Wylie has argued that they are best thought of as a series of tensions "between distance and proximity, observing and inhabiting, eye and land, culture and nature; these tensions animate the landscape concept, make it cogent and productive," and this point—that tension and contention produce a vibrant discipline—does not need to be stressed within the humanities.[61] The rapid development of digital spatial technologies, from mapping to virtual worlds, is sure to increase these tensions as the array of possible representations is greatly expanded. All representations are models, whether maps, diagrams, textual descriptions, databases, GIS applications or VR; they are constructed to simplify some "real world" complexity and to enable thinking and understanding. The central role of interpretation in this process is a new tension when applied to the quantitative basis of digital representation and one that as humanists we are best placed to exploit.

As a final comment, it is interesting to note that space and place have played different roles in the methodologies and theoretical approaches of different humanities disciplines. In geography and archaeology the spatial variable has always been of central interest. Perhaps what we are experiencing now is how new, developing spatial technologies fit into the "humanist turn" rather than the spatial one.

NOTES

1. See Ayers, chapter 1, this volume.
2. John Wylie, *Landscape* (London: Routledge, 2007).
3. Department of the Environment, Handling Geographic Information. *Report to the Secretary of State for the Environment of the Committee of Enquiry into the Handling of*

Geographic Information, chaired by Lord Chorley (London: Department of the Environment, 1987).

4. K. Allen, S. Green and E. Zubrow, eds., *Interpreting Space: GIS and archaeology* (London: Taylor and Francis, 1990).

5. Barbara Bender, ed., *Landscape: Politics and Perspectives* (Oxford: Berg, 1993), 1.

6. C. Sauer, "The morphology of landscape," in C. Sauer, *Land and Life* (Berkeley: University of California Press, 1963), 400.

7. W. Hoskins, *The Making of the English Landscape*, 1st ed. (London: Penguin, 1954).

8. Denis Wood, *The Power of Maps* (London: The Guilford Press, 1992); M. Monmonier, *How to Lie with Maps*, 2nd ed. (Chicago: University of Chicago Press, 1996).

9. In the words of Leonard Cohen: "... and there is no space, but there's left and right, and there is no time, but there's day and night...." "Ballard of the Absent Mare," from *Recent Songs* (Columbia Records, 1979).

10. http://www.commonground.org.uk and in particular Sue Clifford's article, "Places, People and Parish Maps" (accessed 2 May 2008); D. Crouch, and D. Matless, "Refiguring geography: Parish Maps and Common Ground," *Transactions of the Institute of British Geographers*, NS21 (1996), 236–55.

11. http://www.unc.edu/awmc (accessed 2 May 2008); R. J. A. Talbert, *The Barrington Atlas of the Greek and Roman World* (Princeton, N.J.: Princeton University Press, 2000).

12. Gary Lock and Brian Molyneaux eds., *Confronting Scale in Archaeology: Issues of Theory and Practice* (New York: Springer, 2006).

13. Marcia-Anne Dobres, *Technology and Social Agency* (Oxford: Blackwell, 2000).

14. Chris Gosden, *Social Being and Time* (Oxford: Blackwell, 1994).

15. Gail Langran, *Time in Geographic Information Systems* (New York: Taylor and Francis, 1992).

16. G. Fairclough, "Large Scale, Long Duration and Broad Perceptions: Scale Issues in Historic Landscape Characterisation," in G. Lock and B. Molyneaux, eds., *Confronting Scale in Archaeology: Issues of Theory and Practice* (New York: Springer, 2006), 203–16.

17. Fairclough, "Large Scale, Long Duration and Broad Perceptions," 209.

18. W. Kuhn, *Dynamizing Spatial Semantics*, on demand Webcast, The e-Science Institute, Edinburgh: http://www.nesc.ac.uk/esi/esisearch.cfm (accessed 2 May 2008).

19. A. Sheehan-Dean, "Similarity and Difference in the Antebellum North and South," in A. Knowles, ed., *Past Time, Past Place. GIS for History*. (Redlands, Calif.: ESRI Press, 2002), 35–50.

20. P. Haggett and R. Chorley, *Network Analysis in Geography* (London: Edward Arnold, 1969).

21. G. Lock and J. Pouncett, "Network Analysis in Archaeology. Session Introduction: An Introduction to Network Analysis," in J. Clark and E. Hagemeister, eds., *Digital Discovery. Exploring New Frontiers in Human Heritage. Computer Applications in Archaeology Conference 2006, Fargo.* (Budapest: Archaeolingua, 2007), 71–74.

22. L. Symonds and R. Ling "Travelling Beneath Crows: Representing Socio-Geographical Concepts of Time and Travel in Early Medieval England," *Internet Archaeology*, 13 (2002): http://intarch.ac.uk (accessed 5 May 2008).

23. Ian Gregory and Paul Ell, *Historical GIS: Technologies, Methodologies and Scholarship* (Cambridge: Cambridge University Press, 2007). Cited here at chapter 5.

24. Gregory and Ell, *Historical GIS*, 100.

25. P. Daly and G. Lock, "Time, Space and Archaeological Landscapes: Establishing Connections in the First Millennium BC," in M. Goodchild and D. Janelle, eds., *Spatially Integrated Social Science* (Oxford: Oxford University Press, 2004), 349–65.

26. D. Rumsey and M. Williams "Historical Maps in GIS," in A. Knowles, ed., *Past Time, Past Place: GIS for History* (Redlands, Calif.: ESRI Press, 2002), 1–18.

27. R. Witcher, "GIS and Landscapes of Perception," in M. Gillings, D. Mattingly and J. van Dalen, eds., *Geographical Information Systems and Landscape Archaeology* (Oxford: Oxbow Books, 1999), 13–22.

28. B. Devereux, G. Amable, P. Crow and A. Cliff, "The Potential of Airborne LiDAR for Detection of Archaeological Features under Woodland Canopy," *Antiquity*, 79 (2005), 648–60.

29. K. Kvamme, "GIS Algorithms and Their Effects on Regional Archaeological Analysis," in K. Allen, S. Green, and E. Zubrow, eds., *Interpreting Space: GIS and Archaeology* (London: Taylor and Francis, 1990), 112–26.

30. D. Wheatley and M. Gillings, "Vision, Perception and GIS: Developing Enriched Approaches to the Study of Archaeological Visibility," in G. Lock, ed., *Beyond the Map: Archaeology and Spatial Technologies* (Amsterdam: IOS Press, 2000), 1–27.

31. P. Riggs and D. Dean, "Assessing the Level of Visibility of Cultural Objects in Past Landscapes," *Journal of Archaeological Science*, 28 (2007), 1005–14.

32. M. Lake, "MAGICAL Computer Simulation of Mesolithic Foraging on Islay," in S. Mithen, ed., *Hunter-Gatherer Landscape Archaeology: The Southern Hebrides Mesolithic Project, 1988–98: Archaeological Fieldwork on Colonsay, Computer Modelling, Experimental Archaeology, and Final Interpretations* (Cambridge: McDonald Institute, 2000), 465–95.

33. M. E. Hacklay, "Virtual Reality and GIS: Applications, Trends and Directions," in P. Fisher and D. Unwin, eds., *Virtual Reality in Geography* (London: Taylor and Francis, 2002), 47–57.

34. J. Bruck, "Experiencing the Past? The Development of a Phenomenological Archaeology in British Prehistory," *Archaeological Dialogues,* 12 (2005), 1, 45–72.

35. Christopher Tilley, *The Materiality of Stone. Explorations in Landscape Phenomenology* (Oxford: Berg, 2004).

36. B. Olsen, "Material Culture after Text: Re-membering Things," *Norwegian Archaeological Review*, 36 (2003), 2, 87–104. Cited here at 88.

37. Anthony Giddens, *The Constitution of Society* (Cambridge: Polity Press, 1984).

38. Pierre Bourdieu, *Outline of a Theory of Practice* (Cambridge: Cambridge University Press, 1977).

39. Nigel Thrift, *Spatial Formations* (London: Sage, 1996), 37.

40. T. Ingold, "The Temporality of Landscape," *World Archaeology*, 25 (1993), 152–71; T. Ingold, *The Perception of the Environment: Essays on Livelihood, Dwelling and Skill* (London: Routledge, 2000).

41. J. J. Gibson, *The Ecological Approach to Visual Perception* (Boston: Houghton Mifflin, 1979).

42. V. Trifkovic, "Persons and Landscapes: Shifting Scales of Landscape Archaeology," in G. Lock and B. Molyneaux, eds., *Confronting Scale in Archaeology: Issues of Theory and Practice* (New York: Springer, 2006) 257–71.

43. H. Lorimer, "Cultural Geography: The Busyness of Being 'More Than Representational," *Progress in Human Geography*, 29:1 (2005), 83–94.

44. Mike Pearson, *'In comes I': Performance, Memory and Landscape* (Exeter: University of Exeter Press, 2006).

45. Lorimer, "Cultural Geography," 84.

46. Pearson, *'In comes I,'* 16.

47. T. Harris, S. Bergeron and J. Rouse, "Humanities GIS: Adding Place, Spatial Storytelling and Immersive Visualization into the Humanities" (forthcoming).

48. M. Dodge, "Explorations in AlphaWorld: The Geography of 3D Virtual Worlds on the Internet," in P. Fisher and D. Unwin, eds., *Virtual Reality in Geography* (London: Taylor and Francis, 2002), 305–31.

49. http://secondlife.com (accessed 7 May 2008).

50. A. Greaves, *Reconstructing Hadrian's Wall in Second Life* (Glasgow: The Higher Education Academy, 2007): http://www.heacademy.ac.uk/hca/resources/detail/reconstructing_hadrians_wall (accessed 8 May 2008).

51. M. Gillings, "Virtual Archaeologies and the Hyper-Real: Or, What Does It Mean to Describe Something as Virtually-Real?" in P. Fisher and D. Unwin, eds., *Virtual Reality in Geography* (London: Taylor and Francis, 2002), 17–34.

52. D. Marble, "Geographic Information Systems: An Overview," in D. Pequeut and D. Marble, eds., *Introductory Readings in Geographic Information Systems,* (London: Taylor and Francis, 1990), 8–17.

53. The Bodleian Library. Medieval Islamic Views of the Cosmos. The Book of Curiosities (Oxford: The Bodleian Library, 2009). http://cosmos.bodley.ox.ac.uk/hms/home.php?expand=29 (accessed 20 October 2009).

54. http://www.searchmesh.net/default.aspx (accessed 5 May 2008).

55. http://www.timemap.net (accessed 5 May 2008).

56. Charles Leadbeater, *We Think: Mass Innovation, Not Mass Production,* (London: Profile Books, 2008).

57. See the discussion in chapter 8, this volume.

58. http://googlemapsmania.blogspot.com/#top (accessed 10 April 2008).

59. A. Addison, "*A Vedtelen Bolygo or Digital Heritage 2.0: New Directions for a 'Disappearing World',*" keynote speech given to the Computer Applications in Archaeology Conference, Budapest, Hungary, April, 2008.

60. For example see: http://del.icio.us (accessed 11 April 2008).

61. Wylie, *Landscape,* 216.

SEVEN

Mapping Text

MAY YUAN

INTRODUCTION

This essay centers on the idea that innovative semantic syntheses and georeferencing methods can enable transformation of text to maps. Hence, the attempt is to explore the possibilities of projecting text to produce maps and enabling maps to tell stories. Both stories and maps are important and effective frameworks of learning and understanding the world for both children and adults.[1] Humanities Geographic Information Systems (GIS) must incorporate the ability to represent narratives in GIS databases and map texts to offer the geographic contexts of stories. By doing so, we may be able to realize Sir C. P. Snow's vision of "the third culture" that bridges scientific and literary disciplines.[2]

Text is one, if not the, major form that records the human experience with an epistemology of reality. Dumbrava argued that text converts experiences to discourses and enables us to perceive experiences of others and other times.[3] One effective means of communicating experiences in text is through metaphors that bridge the gap between concrete experience and revelation of reality. Consequently, text becomes the space of "culturalization" that transforms history to culture with common ontological and epistemological dimensions of comprehension and sense making.

Maps are also metaphors that can bridge concrete experience and revelation of reality. Maps are one, if not the, most effective form that records knowledge about geography where histories unfold and cultures develop and to serve as a shared reality. As a spatial language, maps apply semiotics to capture geographical ontology and epistemology that address features, patterns, and relationships over space. They present the space of "contextualization" that transforms location to place with syntheses to communicate where things are and how they relate spatially.

Can texts and maps be interchangeable? The linear modes of texts order events chronologically within the extent of inferences to restructure the experience of reality in forms of stories or narratives commonly expressed in oral or written forms for communication, learning, and knowledge building. The intrinsic two-dimensional form of map displays conveys directly the spatial settings and situations in which events developed. Such interchangeability of texts and maps can promote chronologically and spatially integrated cultures and consequently bring about a new dimension of comprehension through the perpetual interplays between cultural and geographical environments.

The reciprocal transformation from texts to maps or from maps to texts exchanges time-centric and space-centric organizations of human experiences. In natural language processing, we understand texts through semantic analysis of multiple dimensions in order to extract meanings and form stories or narratives. These dimensions may include lexicons, grammars, and syntax. In GIS, mapping is through georeferencing that projects features, phenomena, or ideas over space.

There are many conceptual and technical challenges to enable the text-map interchange in GIS. While these conceptual and technical challenges have not yet been fully addressed, researchers in GIS, computer science, linguistics, bioinformatics, and many other groups are exploring new possibilities. This chapter addresses three possible approaches. Outside GIS, mapping texts is usually related to concept mapping and linguistic understanding of texts. Recent developments in text analytics and text mining have been mostly in bioinformatics, business, and social network applications. In GIS, the primary interest is geographic mapping, which is the meaning taken in this chapter, unless noted otherwise. As mapping text is still an exploratory concept in GIS, approaches discussed here are by no means exhaustive. Rather, these approaches are intended to provoke interest and kindle thoughtfulness about interrelating literary and scientific mentalities in order to gain a holistic understanding of how humans live and have lived around the world, as advocated in C. P. Snow's portrait of a third culture.

THREE POSSIBLE APPROACHES TO
TEXT-MAP TRANSFORMATION IN GIS

Text has not been an important source for GIS data. Direct links of text documents from GIS databases are not uncommon, but the utility of text is lacking in geographic database development or GIS analysis. The concern here is not about importing structured text files, such as comma separated values (CSV) files, into a GIS database. In order to use text as a GIS data source, new models and tools are needed to extract values from narratives or stories and to populate a database.

While text is not yet fully utilized as a GIS data source, research to develop text mining tools for database curation is thriving in bioinformatics. For example, an open contest challenged teams to develop systems that automatically determine if papers contain experimental evidence and, if so, to rank papers according to the possibility of the need for curation.[4] Concepts and techniques developed in bioinformatics are worthy of investigation for their applicability to GIS. By converting unstructured text to structured GIS data, we can develop new empirical inquiries and analytical methods to decipher historical and cultural discourses over geography. Below are three possible approaches to convert unstructured text to structured GIS data for mapping and spatial analysis. Spatialization and gazetteer-matching are two approaches that GIScientists have been building as foundations for research and development.[5] Inference through geospatial markers and metaphors brings an additional new thinking to georeferencing stories and narratives.

SPATIALIZATION

The process of spatialization transforms non-geographic data to spatial forms for visual analysis. As a form of information visualization, spatialization has been used for business transactions, news reports, and many other applications.[6] Fundamentally, spatialization reduces data dimensionality by compressing multidimensional variables into two dimensional displays. Many multivariate statistical methods (e.g., principal components analysis and clustering analysis) are simple spatialization techniques. Recent development in self-organizing maps and support

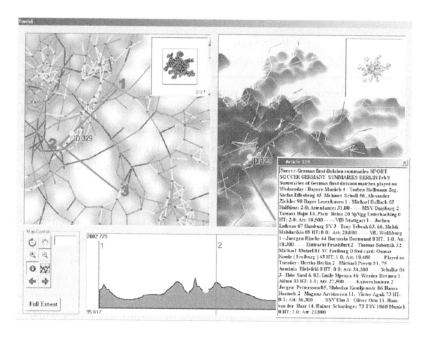

FIGURE 7.1. Spatialized user interface for exploring a news wire archive. All displays here are based on news reports. The shading and the 3-D plot represent the semantic density of the data space. Links among points of news reports indicate the strength (by colors) of cross referencing. The user can draw a transect line to identify implicit cross-references between locations in the data space (locations 1 and 2), and the profile shows the variation of similarity between data points. The user can click on a point (ID 329) and retrieve the corresponding document. From Supkin and Fabrikant 2003.

vector machines apply non-linear mathematics to project data points to their attribute space in two- or three-dimensional displays. By doing so, humans are able to visualize the distribution of data points and to identify clusters or potential groupings of data points based on proximity of these data points in the attribute space. Figure 7.1 shows a system using self-organizing maps, pathfinder network scaling, and spatial interpolation based on a spring model to spatialize a Reuters news wire archive.[7] Unlike geographic maps that show properties of geographic space, all displays in the figure illustrate properties of the data space (or the semantics space). These displays communicate semantic relationships among news reports based on cross-references extracted from text. Locations of these reports

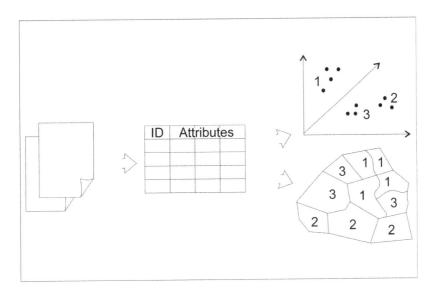

FIGURE 7.2. Spatialization approach to text-map transformation.

are determined by the categorical nature of the reports. For example, news on sports will be clustered and will be close to news on entertainment but distant from news on economy. The actual grouping of these reports depends on their characteristics encoded for analysis.

Coordinates or spatial references in spatialization are attribute-based. Projection from the attribute space to geographic space is often based on known geographic units, such as census tracts or vegetation zones.[8] By comparing clusters in the attribute space and in the geographic space, we can draw insights into potential geographic influences. Figure 7.2 illustrates the approach of spatialization for text mapping. Conversion from unstructured text to structured tables is a non-trivial task. The first challenge is to identify the entities of interest (cf. Figure 7.4). In many cases, entities of multiple levels should be considered, such as individuals, families, and communities. Attributes associated with individuals may or may not be aggregated to families. Emerging properties at a higher level (e.g., a community) may or may not cascade down to members at a lower level (e.g., an individual). Hierarchy theory suggests that a comprehensive understanding of an entity or a process must consider additional entities at the same level, at a higher level, and at a lower level.[9] Entities at different levels may exhibit distinct clustering patterns and therefore suggest singular underlying factors and drivers.

For example, if we have a set of family text documents that describe characteristics for each family and we are able to manually encode these families and their characteristics in a tabular form, then we can apply spatialization technique (such as self-organizing maps) to identify groups of families that share similar characteristics. Moreover, if we know the locations of these families, we can georeference these families by their addresses or places of residency. We can apply spatial clustering techniques (such as K-means methods) to determine family groups in the geographic space.

A comparative study of family grouping in characteristics space and in geographic space can suggest answers to questions of geographic influences. If family grouping in the characteristics space is highly correlated with the grouping in geographic space, then the data suggest that similar families are located in geographic proximity, which implies a degree of spatial autocorrelation. While several indices are available to measure spatial autocorrelation (e.g., Moran's I or Geary's C), the spatialization approach considers multiple variables distinctive of the other methods.

Furthermore, we can also apply spatialization techniques to examine the geographic residencies of these families and then compare clusters of their geographic residencies and their family characteristics. Such a comparison can draw insights into whether similar geographic settings are likely to correlate with similar family characteristics. A significant positive correlation may imply interactions between geographic settings and family characteristics or common factors influencing the developments of families and geographic environments. The use of spatialization for text-map transformation assumes that spatial locations and extents are available for entities identified from the text. Most commonly for census documents, national statistics, or any reports tailored to individuals with addresses or with geographic zones. The following two approaches are for text that does not support such an assumption.

GEO-REFERENCE: PLACE NAME MATCHING THROUGH DIGITAL GAZETTEERS

Gazetteers are collections of place names with geographic locations (or footprints) and other information about these places. The Alexandria

Digital Library (ADL) project at the University of California, Santa Barbara (http://www.alexandria.ucsb.edu/) led the efforts to digitize place names and develop services for accessing place names and linking them to other georeferenced materials as part of the National Geospatial Digital Archive (NGDA) led by the Library of Congress. In addition to preservation, ADL and NGDA creates a cyber-infrastructural foundation for distributed digital libraries of geographic names that are searchable and retrievable online, as well as for digital gazetteer information exchange and knowledge integration. Worldwide, many accomplished projects have extended digital gazetteer networks to ancient place names (e.g., http://icon.stoa.org/pleiades-beta/) and named events in newspapers.[10] In particular, the establishment of content standards for digital gazetteers is instrumental to ensuring the overall usability and interoperability of digital gazetteers.[11] Detailed ADL content standards are available at http://www.alexandria.ucsb.edu/gazetteer/gaz_content_standard.html. In a nutshell, the gazetteer content standard speculates that a comprehensive framework for place names should include three core elements: toponyms (and their history), spatial location (in various representations, such as points, lines, and polygons), and classification (e.g., types and categories of places). Attribution for information from multiple sources also should be noted.

Another significant ADL contribution is to enable "geoparsing" text to identify geographic representation of places and geographic features through matching place names in digital gazetteers. Text geoparsing allows us to link geographic properties of places to place names and enrich our understanding of discourses communicated in the text. From 1994 to 1998, the ADL project organized a series of workshops and produced research reports that made great advances in fundamental development and services applications. These workshops stimulated researchers to develop geoparsing and Web browsing tools. A geospatial browser developed for Los Angeles gazetteer applications highlights the utility of digital gazetteers as an enrichment tool for geospatial neighborhood explorations.[12] In the UK, geoXwalk provides similar ADL services to geoparse academic documents over the Internet.[13]

At the basic level, geoparsing is to extract place names in text through natural language processing. Once the places are identified, we can match the names with entries in a digital gazetteer. Matching place names can be geotagged in eXtensible Markup Language (XML) or Geographic

Markup Language (GML) to GIS services or Keyhole Markup Language (KML) with Google Earth, for example, to provide geographic views with maps and images of these places. There are many forays into geoparsing algorithm design and implementation as presented in workshops and conferences in GIS and computer science related communities, but peer-reviewed journal publications are rare. Most efforts in standardization and tool developments are technology-centric and service-oriented. The lack of rigorous conceptual and methodological frameworks appears to be the key challenge for the academic establishment.

For example, the current digital gazetteers have limited or no support for qualified queries that seek place names based on spatial relationships, and most results are implementation specific to the embedded structure of a digital gazetteer. In addition, geoparsing needs to go beyond simply matching place names or spatial terms. Over time, places may have multiple names, multiple footprints, and multiple identifications that raise questions whether the place remains the same place or should be considered a different place. Place names are often labels synthesizing a coherent space emerging from environmental, political, historical, and cultural interplays. Change in place names or place footprints have profound implications to events occurring in these places. When geoparsing and search results in multiple matches or no match to a place name, algorithms should have the capability to apply auxiliary information in the text to seek rankings or alternatives. While simple geoparsing extracts place names from text, additional semantic analysis of documents is necessary to discover the meaning of text and reveal implications and significance of places to determine proper geographic mapping of these places.

Therefore, in addition to place names, we should consider an integrative georeference framework to decipher geographic knowledge embedded in text and to construct discourses and narratives. Such an integrative geo-reference framework calls for contextualizing place names with events, cultures, and many other environmental and human factors that together make a place idiosyncratic. One of our research projects aims to develop a GIS knowledgebase of world culture systems that include ethnic groups, religions, and languages, along with political boundaries and environmental features. The GIS knowledgebase can contribute to such an integrative framework to extract and map geographic knowledge from text (Figure 7.3). There are many GIS projects on digital cultural atlases (e.g., the Electronic Cultural Atlas Initiative coordinated by the University of California at Berkeley and the Map History as part of WWW-Virtual Library coordinated by British Library). Because it integrates cultural information from multiple books and encyclopedic sources,[14] our emphasis on cultural knowledge and knowledgebase development highlights the needs for spatial fusion and synthesis of cultural manifestations and facts from multiple sources and the desire to develop rules and procedures for analysis and reasoning. By projecting places into a GIS knowledgebase for world culture systems, we can relate the geographic fabric of places and trace interactions among places. Places are integrals of human experiences and should be represented as such. When geographic information is integrated at a place, the place becomes a landmark of cultural ecology.

GEO-INFERENCE

Research on geospatial inference generally centers on identifying spatial rules and applying the rules to make spatial decision, such as site selection. Topological relationships, such as overlaps and disjoint, are commonly used as the basis for spatial queries.[15] A simplistic example is the claim that property values are higher in better school districts, which is a finding obtained by comparing property values against the performance measures of school districts. To make the comparison, we need to determine which properties are located within which school district by overlapping points of property locations and polygons of school districts. If the information is in a text format, or more likely a tabular format, geo-referencing property addresses and school districts must be based on existing geospatial data, such as street networks and municipal boundaries. Once the data are geo-referenced, we can then further investigate interactions of school districts and property values. A spatial pattern of school performances and property values may be discerned, which suggest the underlying interactions of property values and school performance. A district with good schools is likely to drive up property values within the district. Since most public schools in the U.S. are supported by property taxes, higher property values provide better financial support for schools in respective districts, and hence results in an upward spiral cycle of school quality and property values.

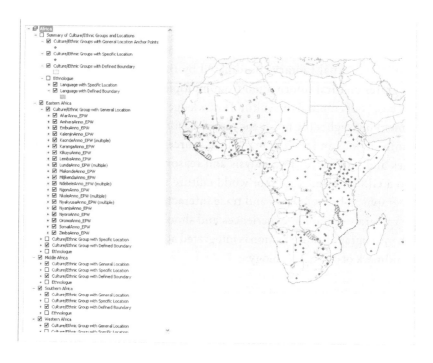

FIGURE 7.3. An example from the GIS knowledgebase world cultures that is under development by the author's research group (Melissa Brown*, Dustin Howard, Daniel Wortham*, Lindsey Wortham*, Eric Burley, Derek Morris). * = graduated in 2007.

The simplistic example illustrates the benefits of spatial inference and analysis by converting text to GIS data. A more interesting approach is to apply spatial rules or inference strategies for text-map transformation. In addition to place names or geographic features names (such as the Grand Teton or the Canadian River), there are many terms that can serve as geospatial markers. The GIS knowledge of world culture systems discussed earlier is an example of how to georeference cultural groups, religions, and languages included in a text document. If a place name has been changed, the combination of geographic features, ethnicity, religions, other information, if available with geographic references, can be used to narrow down the most likely location of the place by conditional reasoning or logic elimination. Furthermore, if the text includes events occurring in the place, an event gazetteer, as discussed earlier, can be used to extract possible matches of the events and chronological orders of these events to reveal broader historical discourses beyond the information embedded in the text. On the other hand, information on events may include places where the events took place. If so, we can georeference these places by matching against digital gazetteers and consequently georeference the events.

Other potential geospatial inferences can be achieved by geospatial markers, such as nearby, at, along, surrounding, meandering, and many other terms that characterize spatial forms, spatial relationships, or spatial proximity in reference to a place or geographic feature. A compilation of geospatial marker sets is important to extract spatial relationships embedded in text. When the location of the place or geographic feature can be resolved based on gazetteers or GIS data, we can then estimate the potential location of interest. If not, semantics analysis of the text may be necessary to understanding its geographic context and through the understanding to estimate potential locations. Similar work is in the area of crime investigation through geographic profiling[16] or animal spotting.[17] An integrated framework that combines place names, feature names, event names, ethnonyms, religions, and other geographic and cultural fabrics is the key to enable a powerful text-map transformation.

There are many proprietary and open-source semantics analysis and text mining tools available. SAS TextMiner, Latent Semantics Analysis, TextAnalyst, YALE text mining, and TerMine are only a few examples. These programs are able to extract terms of interest, visualize these terms, categorize text of similar contents or styles, retrieve relevant text of interest, and many other text analysis functions. Figure 7.4 shows an example of using semantics analysis to convert unstructured news reports to tables.[18] The table can then be mapped to countries in GIS databases to show the spatial distribution of countries that experience disease outbreaks over time. One key to successfully applying any of these tools to humanities GIS applications is to determine terms of interest and contextual relevance. First, we can simply retrieve all nouns from text and attempt to match them against place names and feature names in digital gazetteers and event gazetteers. Secondly, if we have a set of geospatial markers, these text mining tools can apply these markers to reveal spatial relationships among the identified places and furthermore to infer potential location for unmatched places or events.

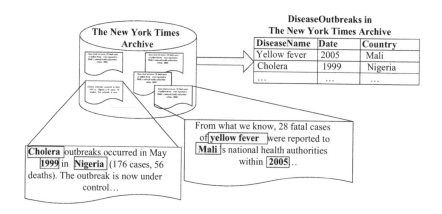

FIGURE 7.4. Building tables from unstructured text through parsing. From Panagiotis et al. 2007.

Once spatial relationships are built among places, we can then use the text mining tools to discern event discourses and attach the discourses to georeferenced places. The next step will be to develop automatic symbolization tools to represent temporal sequences of discourses, such as flows of populations from one place to another by marking an arrow between the two places. The width of the arrow shall be displayed in proportion to the quantity of migrants among places. An expansion of power or a growth of merchant networks may be presented as animations as in TimeMap or GeoTime. Once places and events from text are geo-referenced, spatial analysis and modeling tools can be used to discern previous unknown spatial dependency or patterns that are not obvious in text form. Furthermore, research in the Geospatial Semantic Web will provide a broader basis of text integration with geospatial tagging. Outcome of this research can provide massive information for text mapping to support GIS database curation.

CONCLUDING REMARKS

Text is the richest source of information across almost all subjects. With the ever growing text on the Web, the importance of text as a data source can only increase over time. While document analysis has been used to decipher the information needs for GIS development,[19] the use of text as a GIS data source is still uncommon. This chapter introduces three possible approaches to convert text to maps. Spatialization is an effective means to visualize text and invite exploration, analysis, and interpretation when reading text alone cannot lead to a complete grasp of connections and correlations among documents. Applications of digital gazetteers provide a computerized framework that allows us to develop matching algorithms to automate identification of location references from digital gazetteers. Once georeference is completed, places as described in text can be mapped to geographic space. GIS methods can then be used to analyze spatial patterns and relationships among these places and incorporate additional geographic data to access the broader and more detailed geographic content of the place. Finally, geospatial inferences apply spatial reasoning to estimate places that cannot be recognized through gazetteer search. By interrelating event gazetteers, cultural knowledge bases, and geographic data, we can better determine and understand places. Moreover, there are many text mining tools available that can help automate text extraction and analysis. Building on digital gazetteers, event gazetteers, cultural knowledge bases, GIS data sets, and text mining tools, it is possible to realize a humanistic GIS that takes text as an input and to project places and their interactions from the text to maps. What is still missing is a set of geospatial markers and chronological markers that can relate places and estimate locations of places with no specific names, such as a dugout in Hemphill County on Horse Creek, about four miles northwest of its juncture with the Canadian River. Once places in text can be georeferenced and connected geographically, it is possible to examine these places in semantic space based on their attributes and in geographic space based on their locations. Such a comparison can suggest potential factors and drivers that make a location a place. Another possibility is to use text mining tools and a set of place names to search from documents or over the Web, apply geospatial markers and chronological markers to organize the findings, form narratives about these places, and project narratives to maps to communicate the spatial dimensions of stories.

NOTES

1. Lorna Bennett, "Narrative Methods and Children: Theoretical Explanations and Practice Issues," *Journal of Child and Adolescent Psychiatric Nursing*, 21:1 (2008), 13–23.

2. C. P. Snow, *The Two Cultures and the Scientific Revolution* (Cambridge: Cambridge University Press, 1960).

3. Gabriela Dumbrava, "Turning History into Culture," *European Journal of English Studies* 11:3 (2007), 251–61.

4. Alexander Yeh, Lynette Hirschman, and Alexander Morgan, "Evaluation of Text Data Mining for Database Curation: Lessons Learned from the KDD Challenge Cup," *Bioinformatics* 19 (2003), i331–i339.

5. André Skupin and Sara Fabrikant, "Spatialization Methods: A Cartographic Research Agenda for Non-Geographic Information Visualization," *Cartography and Geographic Information Science* 30:2 (2003), 99–119; Linda Hill, *Georeferencing: The Geographic Association of Information* (Cambridge, Mass.: MIT Press, 2006).

6. Nikolaos Ampazis and Stavros J. A. Perantonis. "LSISOM—A Latent Semantic Indexing Approach to Self-Organizing Maps of Document Collections," *Neural Processing Letter*, 19:2 (2004), 157–73.

7. Skupkin and Fabricant, "Spatialization Methods," 113.

8. Fernando Baçäo, Victor Lobo, and Marco Painho, "The Self-Organizing Map, the Geo-SOM, and Relevant Variants for Geosciences," *Computers and Geosciences* 31:2 (2005), 155–63; Diansheng Guo, Mark Gahegan, Alan MacEachren, and Biliang Zhou, "Multivariate Analysis and Geovisualization with an Integrated Geographic Knowledge Discovery Approach, *Cartography and Geographic Information Science* 32:2 (2005), 113.

9. Valerie Ahl and T. F. H. Allen, *Hierarchy Theory: A Vision, Vocabulary, and Epistemology* (New York: Columbia University Press, 1996).

10. Robert Allen, "A Query Interface for an Event Gazetteers," *Proceedings of the 2004 Joint ACM/IEEE Conference on Digital Libraries* (New York: ACM Publishers).

11. Linda Hill, "Core Elements of Digital Gazetteers: Placenames, Categories, and Footprints," in J. Borbinha and T. Baker, eds., *Proceedings of the Fourth European Conference on Research and Advanced Technology for Digital Libraries* (Berlin: Libraries Unlimited, 2000), 280–90.

12. John Wilson, Christine Lam, and Deborah Holmes-Wong, "A New Method for the Specification of Geographic Footprints in Digital Gazetteers," *Cartography and Geographic Information Science*, 31:4 (2004), 195–207.

13. James Reid, "geoXwalk—a Gazetteer Server and Service for UK Academia," European conference on research and advanced technology for digital libraries, Trondheim, Norway, 2003.

14. Amiram Gonen, ed., *Encyclopedia of the Peoples of the World* (New York: Henry Holt & Company, Incorporated, 1993); D. Levinson, ed., *Encyclopedia of World Cultures*. (Boston: Cengage Gale, 1996); J. Minahan, *Nations without States: A Historical Dictionary of Contemporary National Movements* (Westport, Conn.: Greenwood, 1996); Felipe Fernandez-Armesto, ed., *Guide to the Peoples of Europe* (London: Times Books, 1997); Timothy Gall, ed., *Encyclopedia of Cultures and Daily Life* (New York: Cengage Gale, 1997); Diagram Group, *Encyclopedia of African Peoples* (London: Cengage Gale, 2000).

15. Max Egenhofer and Robert Franzosa, "Point-Set Topological Spatial Relations," *International Journal of Geographical Information Systems* 5:2 (1991), 161–74; M. Egenhofer and R. Franzosa, "On the Equivalence of Topological Relations," *International Journal of Geographical Information Systems* 9:2 (1995), 133–52.

16. Lin Liu and John Eck, "Analysis of Police Vehicle Stops in Cincinnati: A Geographic Perspective," *Geography Research Forum* 27 (2007), 29–51.

17. Steven Le Comber, Paul Racey, Barry Nicholls, and D. Kim Rossmo, "Geographic Profiling and Animal Foraging," *Journal of Theoretical Biology* 240:2 (2006), 233–40.

18. Panagiotis G. Ipeirotis, Eugene Agichtein, Pranay Jain, and Luis Gravano, "Towards a Query Optimizer for Text-Centric Tasks," *ACM Trans. Database Syst.* 32:4 (2007), 21.

19. May Yuan, "Use of Knowledge Acquisition to Build Wildfire Representation in Geographical Information Systems," *International Journal of Geographical Information Science* 11:8 (1997), 723–45.

EIGHT

The Geospatial Semantic Web, Pareto GIS, and the Humanities

TREVOR M. HARRIS, L. JESSE ROUSE,
AND SUSAN BERGERON

INTRODUCTION

The so-called spatial turn in the humanities represents a complexity of ideas and applications and the term is in dire need of unpacking. At the very base level the spatial turn represents an awareness of the significant role that space plays in human actions and events, and specifically the influence that space plays in humanities disciplines. Without question, the spatial turn has been heavily driven by the growing awareness and availability of Geographic Information Systems (GIS).[1] There have long been exchanges between geography and the humanities that extend as far back as Carl Sauer's inaugural work in cultural geography and his use of examples from history, archaeology, and cultural landscape studies.[2] Historical geography has certainly been at the forefront of this symbiosis in seeking to explore geographies of the past through a blend of human geography and the historical method.[3] It is not surprising then that the early usage of GIS in the humanities has been predominantly in historical GIS, which has drawn upon the primary strengths of GIS in the areas of mapping, gazetteers, and vector boundary delineation to support the geographies of major databases such as historical censuses and the production of atlases. Significantly there has been little demonstrated use of GIS in the humanities that draws upon the extensive spatial analytical sophistication of the technology. In the disparate, largely uncoordinated and application-driven foray of GIS into the humanities, mapping has been without question the dominant expression of the geospatial turn and in many ways the (re)discovery of the power of the map has become synonymous with the spatial turn.

The temporal lag in the uptake of GIS in the humanities is in stark contrast to the rapid diffusion and substantive inroads that GIS made in the social sciences and the sciences over the past three decades or more. The humanities emphasis on GIS as a mapping tool and the dearth of spatial analytic studies is both reflective of the nature of humanities enquiry themselves and the characteristics and qualities of GIS technology. To date, the spatial turn has been an exchange between a relatively few number of scholars in the humanities and in the GIS community that has been predominantly method-driven. The substantive questions about the nature of the spatial turn itself and the forms that the "turn" will or should take must be posited on fundamental ontological and epistemological issues that go beyond the project-driven use of GIS per se. These exchanges must extend well beyond the technology and embrace the core linkages between geographical concepts and humanities enquiry. While the term humanities GIS broadens the focus of investigation beyond more delimited terms such as historical GIS, a focus on humanities Geographic Information Science (GISci), as opposed to humanities GIS as system, better represents not just the geospatial technology involved in the spatial turn but also the conceptual breadth and depth associated with the dyadic interface of a spatial information science with many other disciplinary domains. The use of GIS as a lens to understand the geographical dimensions of the humanities raises questions about the biases, assumptions, and the silences in the technology that impinge upon the exploration of the spatial turn. It also raises questions for the spatial turn for those areas that are not so easily explored through the medium of GIS. For the humanities to focus solely on spatial location to the omission of broader geographical concepts is like asking historians to focus solely on dates to the omission of the deeper knowledge the discipline brings to understanding human nature, behavior, and events that lie at the core of history. If GIS is indeed the early focal point of exchange and dialog around which discourse on the spatial turn revolves then it is reasonable to question what substantive theoretical developments have arisen or may arise from this exchange and to question the constraints that GIS imposes as a way of knowing on the humanities. The GIS and Society debate in the 1990s brought these issues front and center within geography and many of the themes raised in that discourse are equally applicable here with regard to the humanities and GIS.

GIS as a system has evolved to its present form based on an extensive social history that has yet to be fully told.[4] The technology is characterized by a focus on location and the representation of features through the spatial primitives of points, lines, polygons, pixels, and the mathematical geometry or topology that underpin the data manipulation and integration. The significance of measured location, spatial relationships, and attributes heavily underpins the geospatial method as demonstrated by GIS. Spatial data is central to GIS and provides the foundation for spatial analysis and deriving critical information about geographic space. Despite the focus on GIS as a core component of the spatial turn in the humanities it is ironic that the technology is somewhat at odds with many of the ways in which traditional humanities scholarship and methods are practiced. The early focus on GIS mapping has obscured this conundrum and the field has not yet addressed the more fundamental issues concerning the way in which the humanities could or should embrace GIS and, to a greater extent, geography. The importance of place as opposed to space; the need to address qualitative as well as quantitative aspects of place; the need to be empathetic to the story telling tradition in the humanities; the significance given to the linear narrative as represented by the text book which is in contrast to the non-linear digital world of GIS; the need to provide a spatial storytelling equivalent; and the importance of maintaining interpretive ambiguity so prevalent in the humanities against the objective solution space of GIS, are all highly problematic when addressing GIS uses in the humanities. As reflected in the adoption lag of applying GIS in the humanities, the ways of knowing the world as practiced through GIS and the spatial analytic method are not necessarily intuitive or sympathetic to traditional humanities scholarship.

GIS then, encompasses certain expectations, assumptions, traditions, methodologies, and functionality that, while it has been found to be a valuable tool in early applications of GIS in the humanities, are nonetheless in practice an expert technology fashioned in the worlds of science and social science. As an expert technology the science is limited to those with access to the hardware, software, and the humanware needed to effectively utilize its capability. More recently, there has become available a new suite of technologies that, as Andrew Turner suggests,[5] threatens the hegemony of GIS. Within the past three years there has been a revolution in the role of Web services in supporting spatial data delivery and Web-based data integration applications in the form of mashups.[6] These spatial delivery systems are built on the second generation of Web applications and enabling technologies; the so-called Web 2.0. The Geospatial Web is built on interactive Web applications that draw on personal and external data sources to create new innovative geospatial services. In this paper we suggest that the freely available Geospatial Semantic Web represents a Pareto GIS that more closely meets the spatial needs of humanities scholars. More specifically we suggest that by enabling data producers and consumers to share the same technology the Geospatial Semantic Web potentially forms the core of what constitutes a humanities GIS.

NEOGEOGRAPHY AND PARETO GIS

In very recent years there have been significant developments in the Geoweb arising from a new phase in Web development.[7] The Geoweb revolution is best recognized from the geospatial and mapping products of Microsoft, Google, Yahoo, and particularly Web 2.0 geobrowsers. These developments take the form of virtual globes (as with Google Earth, Virtual Earth, and Worldwind) and, since 2005, the availability of Web-based Application Programming Interfaces (APIs), User Created Content (UCC), and Web services. These free Web mapping applications have moved map production into a completely new arena. Michael Goodchild has proposed Volunteered Geographic Information that build on a merger of new Web services and digital sources and geotagged Web entries (e.g., Wikipedia), place descriptions (Wikimapia), public domain map layers (e.g., Flickr); and mashups (e.g., Google Earth and Google Maps). Neogeography is a term sometimes attached to technologies that empower general users to handle geospatial data without having to grapple with the often complex systems encountered in the use of GIS. The first text also has appeared with Arno Scharl and Karl Tochtermann's, *The Geospatial Web: How Geobrowsers, Social Software and the Web 2.0 are Shaping the Network Society* (2007).

In the mid-1990s and amid the burgeoning use of GIS, the GIS and Society debate that was predominantly limited to geography grappled with the issues surrounding the perceived biases embedded in the use of GIS in representing nature and society. This social-theoretic critique questioned, perhaps for the first time, the important implications of the widespread use of GIS.[8] This discourse provided a valuable framework with which to explore issues surrounding the assumptions and biases of GIS.[9] The subsequently evolving fields of Critical GIS and Participatory GIS (PGIS) focused heavily on community mapping, critical counter mapping, and community empowerment and questioned the assumptions that are implicitly embedded in the technology and particularly with regard to how society is represented within the GIS abstraction of reality. This digital abstraction occurs through the use of spatial primitives, structural knowledge distortion, representation, the political economy of GIS, and the logic systems imposed by the technology. In using GIS, certain ways of knowing are privileged over those of others. The similarities between the GIS and Society debate and the issues surrounding the spatial turn in the humanities and the use of GIS in that process bear remarkable similarity. While the GIS and Society debate focused more on the critique of GIS than on the development of tools to grapple with those issues, nonetheless alternative forms of GIS were explored that were designed to be more robust and resistant to the assumptions implicit in traditional GIS. Versions of a GIS 2.0 were thus envisaged that sought "richer" outcomes involving local knowledge production and multimedia GIS—a knowledge-creation environment that surpassed traditional GIS by drawing on many contributors and differing forms of data representation. Thus GIS 2.0 would emphasize the role of participants as data producers and consumers; would accommodate equitable representations of diverse views; would preserve contradictions against premature resolution; would create outputs that would reflect the standards and goals of the participants rather than the closeness of fit to technical standards; that would be capable of managing and integrating all data and participant contributions using one interface including sketch maps, narrative, text, images; would be able to handle time components; would preserve the history of its own development; and would place complementary emphasis on both space and place.

Despite this extensive wish list, Participatory GIS evolved into essentially a GIS mapping process where the more advanced spatial analysis functionalities of GIS were rarely utilized. Much of what constitutes PGIS could be achieved using simpler digital mapping software. This is not to deny the underlying power and utility of GIS, merely its dominant use as a mapping system as employed by the non-expert end-user. The similarities between PGIS and the use of GIS in the humanities is stark and worthy of note. In the 1990s the Internet was almost immediately rejected by PGIS experts as not providing the needed credentials for a GIS 2.0. In the light of the recent evolution in the Internet technologies and Web 2.0 that assertion is now unsupported. The recent flurry of interest in Volunteered Geographic Information reflects this revised thinking. With the emergence of the Internet built on Service Oriented Architecture (SOA), Web Services are at the heart of the new Web 2.0 growth and it is apposite that humanities computing should be ready to take advantage of these new technologies. Through the linkage of geo-spatial and Web services the potential exists to do more than transition digital humanities along an incremental pathway in GIS software adoption but to leapfrog into a convergence of humanities computing, digital humanities, and humanities GIS. We suggest that the Geospatial Semantic Web, built on a combination of GIS spatial functionality and the emerging technologies of the Semantic Web, is capable of providing the core of a humanities GIS able to integrate, synthesize, and display humanities and spatial data through one simple and ubiquitous Web interface. The geospatial server platform would serve to integrate, display, and analyze humanities data in a spatially-enabled format and query and serve data through a semantically-linked Web interface. The potential for humanities scholars to leverage semantically-enabled Web services from a variety of disparate sources, to integrate these sources within the GIS client, to conduct analyses, and to display and disseminate the results via Web Services is phenomenal.

Using the Geospatial Semantic Web, humanities scholars become both data producers and consumers and the Web interface becomes a powerful spatial enabling environment that is arguably more suited to the needs of the humanities than the focus on existing GIS technology. In situating the Geospatial Semantic Web as the cornerstone of humanities GIS we are essentially proposing a Pareto GIS. The Pareto principle, also known as the 80:20 rule, is premised on the law of the vital few and the principle of factor sparsity. Essentially it states that for most events, 80 percent of the effects come from 20 percent of the causes. The term was coined after the Italian economist Vilfredo Pareto who determined that 80 percent of the income in Italy went to 20 percent of the population. Whether conceived as a rule of thumb or a pseudo-scientific law, this non-exact rule can be seen to apply to many different things ranging from wearing 20 percent of our clothes 80 percent of the time to selling 80 percent of a firm's products to 20 of its customers. We suggest that the term is particularly suited to use in the context of a humanities GIS where a critical assessment suggests that only 20 percent of GIS functionality is sufficient to garner 80 percent of the geospatial-humanities benefits. We suggest that while the Geospatial Semantic Web is no replacement for a full-fledged GIS, it is nonetheless a Pareto GIS with substantial benefits that is more than capable of meeting the vast majority of humanities GIS needs. Indeed, given the non-exact nature of the Pareto rule, it could be argued that the proportional divide might even be greater.

THE REVOLUTION OF THE GEOSPATIAL WEB

Mapping and cartographic representation has been embraced by the humanities as the primary method of conveying spatial information. The visual display of information creates a visceral connection to the content that goes beyond what is possible through traditional text documents. Maps, through their visual nature, provide a context for information that allows users to gain a holistic understanding of geographical and spatial information. Advances in GIS and digital cartography now allow enriched content to be attached to traditional map data through the use of spatial multimedia and the integration of qualitative information into the natively quantitative GIS thereby enriching and contextualizing space. Whether through the implementation of a multimedia GIS or by embedding or geotagging multimedia information based on its relative location, the ability to move beyond the static information that is representative of traditional cartography is important to the humanities.

The Geospatial Web builds on these concepts of representation, visual display, and spatial multimedia through the extension of the map using Web 2.0 technologies that increase the utility of cartographic products significantly through user interaction and the dynamic nature of their makeup. The evolution of Web 2.0 has brought with it two distinct trends. The primary trend has been toward social interaction through user generated content, followed closely by a shift in the technology used to build Web content. Social interaction has been fostered through the explosion of wikis, blogs, and other new media. In the Geospatial Web this has largely been based on location-based or geotagged information created by users. The technology used to build Web content has shifted toward a dynamic user experience that uses asynchronous technologies to reduce the load times for content to create a seamless user experience. The best example of the use of such asynchronous technologies are Web mapping applications such as Google Maps and Microsoft's Virtual Earth that allow for user-map interaction that is smooth and considerably more responsive than early era Web mapping technologies.

While the term Geospatial Web is often used, it is the user generated content combined with more traditional spatial data that has driven the growth in this area. User generated content might be accessed through the Web, mobile devices such as cell phones or PDAs, or desktop applications. With the ubiquity of cellular phones it is clear that location-based services will play an ever increasing role in the creation and use of geographically located information. Mobile device technology, however, is still limited due to processor, operating system and size constraints. In contrast, desktop hardware has exceeded many users' needs making it possible to take advantage of graphic intensive applications such as virtual globes which are now being used to great effect to present an array of spatialized content including humanities information. For example, Google Earth's Global Awareness layer contains themed multimedia content related to the humanitarian crisis in Darfur and includes photos, video, audio accounts, and links to Web sites and other resources.

One of the most compelling aspects of these technologies, whether Web, mobile, or desktop, is their use as collaboratory space. The ability to share location relevant information and data within location aware software between interested individuals and groups is robust, but the capability to add additional information, contribute thoughts and comments regarding the information, even to create related media to an initial piece of content allows for a dialectic that can transcend time, space, and community. Many of these functions were core elements in Participatory GIS, which developed in the 1990s as a mechanism for addressing issues related to expert-driven, top-down knowledge structures of existing GIS. However, even PGIS projects were still tied to the political economy of expensive GIS software packages requiring advanced knowledge and extensive resources such that community participation was still required to be mediated through experts in order to build and distribute the GIS and its mapping products. In some cases, Web mapping applications were developed to broaden access further, but these also required support from experts to build and maintain.

The release of Google Maps in 2005 fundamentally changed the landscape of Web mapping. This free Web mapping service took advantage of new Web technologies and especially Asynchronous Javascript And XML (AJAX) to offer a lightweight, fast mapping platform with base vector data and imagery. More importantly, the release of an Application Programming Interface (API) meant that non-expert users could now map their own datasets onto Google Maps base data layers. These mashups could be displayed on any Web site, and combined with other Web technologies to create elaborate collaborative spaces where virtual communities could view, comment, and contribute their own information to the mashup. In addition, projects such as Platial offered a platform from which to gather and coordinate user-generated datasets as well as tools for non-expert users to more easily build their own Web sites.

Google Earth, released in the fall of 2005, built upon the phenomenon begun with Google Maps, but added the novelty of a virtual globe and the ability to explore data in a pseudo 3-D environment. Although Google Earth is essentially a lightweight desktop application, its base data layers and added content are served via streaming Web services that allow for real-time update and access to datasets from numerous sources. In addition, Google Earth and Google Maps support embedded multimedia such as photographs, text, oral narrative, sketches, video, and audio within the map or globe representation thereby allowing users and communities to upload and share spatialized qualitative information that often provides unique insight into aspects of place.

Through its acquisition of @Last Software's SketchUp in early 2006, Google now also supports the display of rendered 3-D models of structures in the Google Earth environment. By using the free Google SketchUp application and Google's 3D Warehouse, any user is able to create, display, and share models of real-world objects, including buildings, vehicles, and trees. These models are easily uploaded to Google Earth and can be utilized to create virtual reconstructions of landscapes anywhere on the globe. Increasingly, these tools are utilized for humanities applications and especially in the modeling and reconstruction of historical structures and landscapes such as the National Register of Historic Places collection at Google 3D Warehouse.

THE GEOSPATIAL SEMANTIC WEB: BEYOND THE SPATIAL COORDINATE

While the emerging Geospatial Web has had a significant impact on the way users can contribute data, create applications, and interact with the information available via the Internet, the development of mechanisms for facilitating the discovery, location, and connection to that data is now a primary focus of research. In 2001, Tim Berners-Lee, one of the architects of the World Wide Web, presented a vision for next generation interaction through the Internet, the Semantic Web.[10] He envisioned the semantic Web as an extension of the current World Wide Web, where content could be shared between computers throughout the Internet because each bit of information had a well-defined meaning. While the use of metadata and standards of interoperability already provide the ability to search within, and between, Web sites to locate information through keyword search engines, such tools do not have the capability to define or interpret the cultural, spatial, or semantic meaning of the search terms or results.

Although research into creating a semantic Web is still in its early stages, a number of technologies have been developed and are being used today, including eXtensible Markup Language (XML) and the Resource Description Framework (RDF). XML allows users to create tags for pieces of data that give structure to that information, but not its meaning. RDF addresses use tags to assign properties and values to certain items and thereby create relationships. Although XML and RDF are not the only solutions so far proposed in the development of a semantic Web, they are useful in helping to define and represent how structure might be added to the content of the World Wide Web and allow computers to perform many of the logical reasoning tasks involved in finding and interpreting appropriate information from the vast amounts of data already available through the Internet.

We suggest that the notion of a semantic Web relates closely to the needs of a humanities GIS. As with other knowledge domains, there is a wealth of content available for researchers in the humanities, but the majority of that content does not lend itself to the traditional quantitative structure of GIS data. Integrating humanistic data into a GIS platform has focused on mapping individual items, such as digital photographs or other multimedia, to a location on a map. While such contextual GIS provide useful syntheses of disparate data sources, they are not linked to other repositories of humanities data in any meaningful way and are thus limited to the data sources available to each project. Recently, separate initiatives within GISci and Digital Humanities have begun to converge around the goal of developing a semantic Web of information connected by the World Wide Web. It has been necessary to find ways in GIS to analyze and categorize the qualitative information that is so central to humanities scholarship. The goal of a Geospatial Semantic Web is not only to semantically decipher the meaning of appropriate phenomena stored on the WWW but to automatically identify the spatial location of phenomena whether mentioned within a text or other representation and to then assign that information to a map. In this way the Geospatial Semantic Web goes beyond an emphasis solely on spatial coordinates to include places based on description, relative location, and even visual representation such as sketches and photographs. For example one might want to handle the linguistic differences between a historic town name in a book and its modern placename counterpart. The two may coincide in geographic space, but there is no easy way to connect the two when geocoding the place names. The resolution to this issue is to use semantic libraries that automatically provide the critical relationship between the different names.

Semantic libraries act like a Rosetta Stone positioned between related descriptions in texts or between the very large and rich datasets that have been created through different efforts. Many humanities databases have been created in response to individual needs with little or no interaction with groups conducting similar research or with similar data collection needs. However, with the move toward open data sharing and service oriented architectures (SOA) there has been a realization that there needs to be a way to work between data sources, to join data, or to even merge data sets. The formalization of semantic relationships into ontological libraries allows data to be linked using standardized semantic relationships. The result of these linkages between data sources and the interpretations of text descriptions allows for more to be done by the computer and removing much of the manual work to be performed by the end user. These formalized ontologies or taxonomies represent a much needed layer to ensure interoperability between data sources and allow researchers interested in creating ontologies the ability to support open, standardized ontological libraries and thereby remove the need to recreate customized semantic relationships for every project.

Outside of the creation and implementation of domain-specific ontologies, semantics allow us to create software that can parse text documents for various forms of analysis. Geoparsing is most often used to find specific place names, named physical and cultural features, and regions. However, geoparsing technologies can also take advantage of descriptive text such as "near," "to the east of," or other descriptions of location. This ability allows us to extend past the idea of location in texts as gazetteers to use maps in the humanities to show locations, relationships, and act as a context for content of information sources. The spatialization of information from individual texts to linked databases on the Internet can be shared through the use of the most notable concept of the Web 2.0 movement, the mashup.

A GEOSPATIAL WEB HUMANITIES CASE STUDY

There is a wealth of Geospatial Web technologies that have existed since the mid-1990s including ESRI's first attempt at an Internet mapping system, MapObjects IMS, through to Mapquest.com, one of the earliest Web mapping sites. These Web 1.0 technologies have given rise to Geospatial Web technologies that draw upon available robust technologies and enables users to draw upon existing and user generated data to create new, user friendly, Internet-based geospatial tools. Humanities scholars can utilize Geospatial Web tools and applications to develop collaborative, interactive mapping environments that incorporate traditional geospatial data with existing multimedia and new media components, and newly created user-generated content. As an example, we have developed in a very short space of time a Web mapping project using new Web mapping and virtual globe technologies that combines spatial data information, such as polygon footprints of historic structures, with embedded multimedia, including historical photographs, audio commentary, and text: information that is typical of the breadth of a humanities project application.

The case study for this research, Morgantown, West Virginia, is a small regional city on the Monongahela River that experienced rapid urban growth in the early twentieth century due to the expanding coal industry, lead glass manufacture, the coming of the railroad, and related commercial activities. These industries led to an influx of migrant workers which spurred a building boom in the present day downtown area of Morgantown, especially during the interwar period. Morgantown's economic boom, however, was short-lived as the industrial base that had fueled the area's prosperity declined and led to increasingly difficult economic conditions. By the 1980s, many of the commercial and industrial buildings constructed during the first half of the twentieth century had been abandoned. Within the past twenty years urban redevelopment, gentrification, and economic revitalization have led to the demolition or conversion of many of these structures. These profound changes in the downtown Morgantown area have fundamentally altered the urban landscape and it is difficult to appreciate the historic landscape and sense of place that still evokes memories in long term residents of the city. The historic downtown project sought to demonstrate the ease and power of creating a Geospatial Web application that used the map and 3-D scene as a portal to a wealth of spatial and archival digital historical data; to tie this data to 3-D building structures embedded in the scene; and allow local inhabitants to add their own local knowledge and multimedia resources. This resource was created in a matter of days, with minimal expertise, and allowed data producers and consumers to share the same technological space. As a Pareto GIS it is both different from a "traditional" GIS and yet in many ways more powerful and better suited to humanities needs than a map coverage so representative of current humanities GIS applications.

To develop the Web mapping portion of the Morgantown case study, a base map was created using the Google Maps free API, and significant locations in the downtown area at the turn of the twentieth century were tagged with basic information. Historical photographs and multiple media data were also embedded and tagged at specific locations. Although this task is relatively difficult within traditional GIS, adding and editing information within Google Maps requires no expert knowledge or com-

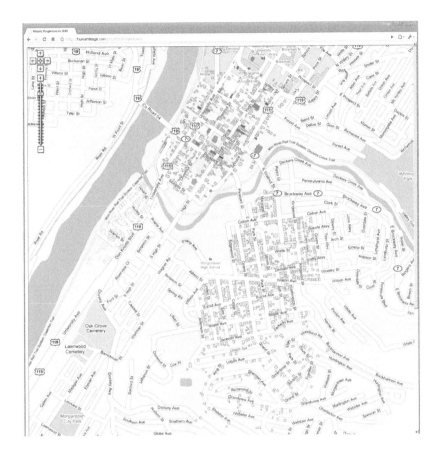

FIGURE 8.1. KML overlay.

putational expertise. Not only do free Web mapping applications allow users to demarcate point locations for text and multimedia tags, these systems also allow more experienced users to add GIS data sets such as point, line, and polygon layers. The primary historical resource utilized for generating the GIS data layers for the virtual reconstruction of early twentieth-century Morgantown consisted of Sanborn fire insurance maps published in 1899 and 1904, as well as historical photographs. The building footprints, street outlines and lot boundaries were all digitized into a standard GIS data layer, then exported into a KML open data format. The resulting KML file was used as an overlay when creating the Google

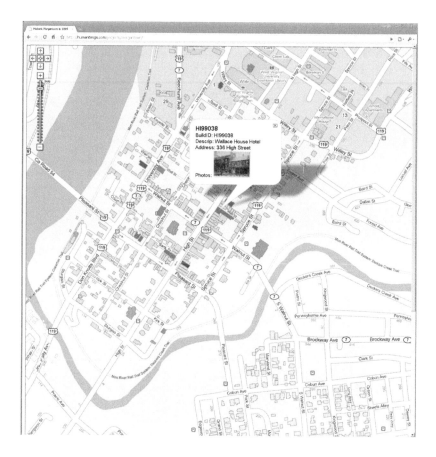

FIGURE 8.2. Embedded media and attribution.

Maps project (Figure 8.1), allowing the easy integration of traditional GIS vector data and multimedia layers in a single application.

One of the most compelling features of the new generation of Web mapping applications is the ability to effectively share the map and allow multiple users to contribute data. For example, the case study of early twentieth-century Morgantown includes entries from students and members of the community, who have added photographs and videos of historical buildings, along with audio discussing aspects of Morgantown's historical landscape (Figure 8.2). While formalized GIS data is an important way to capture and convey data about a landscape, the ability to integrate user generated content provides a tremendous capability for

FIGURE 8.3. Data viewed on a virtual globe.

enriching data content and gaining a sense of place and much more so than with the spatial primitives offered by the GIS alone. The addition of spatially relevant multimedia content such as textual descriptions, pictures, audio, and video extends a user's connection to the place represented by the Web map.

The focus on data interoperability formats with the Geospatial Web arena has resulted in the facility to utilize data in different Web and desktop applications. The KML data format, for instance, is portable to an ever increasing range of applications including Microsoft's Virtual Earth, ESRI's ArcGIS Explorer, and the progenitor of the KML format, Google Earth. The data layers created for the Google Maps project can be viewed in the desktop virtual globe application Google Earth, and, by adding 3-D symbology to the data, a new dimension can be given to the landscape to generate a basic virtual reconstruction (Figure 8.3).

CONCLUSION: WHAT'S IN A NAME?

In this essay we suggest that GIS has played a significant role in raising the profile of the spatial turn in the humanities. We suggest that the spatial turn has been fixated on method rather than on the deeper epistemological and ontological issues associated with linking geospatial technologies and geographical concepts with the needs of the humanities. The early focus on GIS mapping, atlases, gazetteers, and the creation of historical boundary files reflects both the strengths of GIS and the low hanging fruit associated with GIS usage in the humanities. Substantial challenges lie in wait as the full implications of the spatial turn unfold. We suggest that the Geospatial Semantic Web provides a unique opportunity to pursue the spatial turn as a Pareto GIS capable of broadening access for humanities scholars to geospatial technologies. The availability of the Geospatial Web to perform much of the functionality required by humanities scholars is already impressive. Accessibility, ease of use, base mapping, 3-D scenes, point, line, polygon integration, and the power provided by API capability, along with semantic interpretation, promise significant rewards. Importantly, the Geospatial Web closes the very wide gap that currently exists between data producers and data consumers in the humanities. The Geospatial Web, whether accessed via an Internet browser or desktop application, provides access to both spatial data and attribute information. Different services and researchers are working on ways to capture and categorize spatial information that is available online; however there is already a wealth of textual information on the Web that contains spatial information that is ripe for use. By using one of the existing geoparsing technologies, such as MetaCarta's commercial and open APIs, it is possible to review humanities texts for content that can be spatially attributed, and for news streams to be scoured for geographic feature names and relative locations. Perhaps the best example of geoparsing is Jules Verne's *Around the World in Eighty Days* which follows Phileas Fogg through his whirlwind journey around the globe.

The case for neogeography and for access to tools from outside traditional GIS is a powerful one. The perceived hegemonic and throttling grip of a scientifically based and expertised GIS technology may indeed be one contributory reason for the slow uptake of the technology in non-expert and non-scientific communities. However, the power of the Geospatial Web to create and share maps on the users' terms, to convey context and understanding through place, and to perform spatial story telling through contextualized mapping is compelling. In the same way that Participatory GIS has become an important component of contemporary GIS, the parallel lessons for humanities GIS are intriguing to ponder. Volunteered Geographic Information based on the Web 2.0 promises complementary lines of enquiry and possibilities. Contributory humanities GIS using the Geospatial Semantic Web would provide a powerful addition to the geospatial armory for humanities scholars to grapple with the substantive issues of space and place and to engage fully with underlying geographical concepts. The Geospatial Web may have an impact on GIS and map production similar to that of desktop computing on western society. The Geospatial Semantic Web may indeed become the cornerstone of the future humanities GIS.

NOTES

1. David Bodenhamer, Etan Diamond, and Kevin Mickey, *Mapping the Mainline: Using Historical GIS to Study American Religion* (ECAI e-Publication with California Digital Library) http://ecai.org/epubs/Nara/nara_introduction.html (accessed 14 Oct. 2009); Ian N. Gregory and Paul S. Ell, *Historical GIS: Technologies, Methodologies and Scholarship* (Cambridge: Cambridge University Press, 2008); Trevor M. Harris, "GIS in Archaeology," in Anne Kelly Knowles, ed., *Past Time, Past Place: GIS for History* (Redlands, Calif.: ESRI Press, 2002); Anne Kelly Knowles, "Emerging Trends in Historical GIS," *Journal of Historical Geography* (2005), 33; Anne Kelly Knowles, ed., *Placing History: How Maps, Spatial Data, and GIS are Changing Historical Scholarship* (Redlands, Calif.: ESRI Press, 2007); J. B. Owens, "What Historians Want from GIS, *ArcNews*, 29:2 (2007), 4–6.

2. Carl Ortwin Sauer, "The Morphology of Landscape," reprinted in John Leighly, ed., *Land and Life: Selections from the Writings of Carl Ortwin Sauer* (Berkeley: University of California Press, 1963), 315–50.

3. Alan R. H. Baker, *Geography and History: Bridging the Divide* (Cambridge: Cambridge University Press, 2003); D. Clayton, "Historical Geography," in Ron J. Johnston, Derek Gregory, Geraldine Pratt, Michael Watts, eds., *The Dictionary of Human Geography* (Oxford: Blackwell, 2000), 337–41.

4. Timothy W. Foresman, ed., *The History of Geographic Information Systems: Perspectives from the Pioneers* (Upper Saddle River, N.J.: Prentice Hall, 1998).

5. A. Turner, *Introduction to Neogeography* (Sebastapol, Calif.: O'Reilly Press, 2006).

6. Duane Merrill, "Mashups; The New Breed of Web App," http://www.ibm.com/developerworks/xml/library/x-mashups.html (accessed 10 Feb. 2010).

7. L. Jesse Rouse, Susan Bergeron, and Trevor M. Harris, "Participating in the Geospatial Web: Collaborative Mapping, Social Networks and Participatory GIS," in Arno Scharl and Klauss Tochtermann, eds., *The Geospatial Web: How Geobrowsers, Social Software and the Web 2.0 are Shaping the Network Society* (London: Springer-Verlag, 2007), 153–58; "The Spatial Web: An Open GIS Consortium (OGC) White Paper," http://www.openGIS.org (accessed 10 Feb. 2010; Declan Butler, "Virtual Globes: The Web-Wide World," *Nature*, 439 (2006), 776–78; M. Egenhofer, "Toward the Semantic Geospatial Web," *GIS* 2002 (ACM).

8. John Pickles, *A History of Spaces: Cartographic Reasoning, Mapping, and the Geo-Coded World* (New York: Routledge, 2003).

9. Trevor M. Harris and Daniel Weiner, "Empowerment, Marginalization and Community-Integrated GIS," *Cartography and Geographic Information Systems*, 25:2 (1998), 67–76; Nadine Schuurman, "Critical GIS: Theorizing an Emerging Discipline," *Cartographica*, 36:4 (1999), 1–108.

10. T. Berners-Lee, J. Hendler, and O. Lassila, "The Semantic Web," *Scientific American*, 284 (2001), 5.

NINE

GIS, e-Science, and the Humanities Grid

PAUL S. ELL

INTRODUCTION

The development of electronic resources for use by scholars in the humanities has proliferated at a dramatic pace over the last twenty years. Although scholars might feel that few resources are available to them, this is likely not to be the case. Much effort, and funding has been devoted specifically to create e-resources, ranging from highly specialized and subject-specific material to, and of more import to most scholars, what might be termed strategic or key resources. These latter resources might be considered strategic because of their spatial spread (i.e., they provide information for a spatially large area), their spatial granularity (providing information at a detailed spatial level), their chronological depth (data available over long time-periods), or their contextual nature. They are consulted and used by relatively large numbers of scholars, forming, if not a core foundation for their research, at least a backdrop. Such e-resources vary in their nature and include national censuses, socio-economic surveys, the work of mapping agencies, thematic collections of monographs, manuscripts, and journals, and so on.

Many of these resources, perhaps most of them, are primary quantitative sources, especially historical and modern census statistics. Now, for almost all countries with a historical tradition of population censuses, at least part of these data are available in machine-readable form. It should be noted that census-based data is likely to have wider appeal than might be at first apparent. The interests of census enumerators ranged far from a simple count of the population and related demographic statistics. Information might include data on religious adherence, literacy, languages spoken, occupation, economic status, housing and more. Other statistical e-resources include, in Europe, a range of medieval and early modern tax assessments. The most notable of these assessments are various transcriptions and extracted data from the English *Domesday Book* of 1086 with many other subsequent revenue-related surveys following. More recent examples include electoral statistics, agricultural statistics, tithe surveys, and much more.

Increasingly many text-based e-resources have become available. These texts often reflect more closely the interests of the majority of humanists. Of particular significance are transcriptions of previously difficult-to-access manuscript materials such as diaries of key historical and near-contemporary figures, prosopographies, records of debates in senates and parliaments, and travelers' tales. The utilization of these materials on a grand scale affords even more challenges than the use of statistical data. Quantitative information tends to have an internal structure that lends itself to systematic interrogation, even though that structure may be complex and contradictory, as scholars of the Domesday Survey surely would concur. Text-based material, however, can be largely unstructured, thus making the use of these materials in a systematic way difficult.

Currently humanities GIS, still in an embryonic state, tends to be limited to certain disciplines, such as Historical Geography, is largely quantitatively based, analyzing census data, for example, and focuses primarily on maps and other visualizations. Humanities GIS conceived in these terms will always have limited appeal. Most humanists do not deal with quantitative data, come from subject areas far away from geography, and have little interest in visualizing information through maps, time-series visualizations, or alternative/recreated realities. Given such disinterest, is GIS in the humanities always destined to be a part-player in terms of ICT-driven methodological innovations which will shape research in the future?

As currently practiced, GIS will remain a backwater for humanities scholars even if they master the challenges of learning and using the software, much less move to the advanced stages described elsewhere in this volume. GIS has the potential, however, to become a widely used core methodology in the humanities. This key role for humanities GIS does not involve quantitative data alone, and most radically does not require the scholar to have any interest in the spatial nature of their data. Nor will its primary output be maps or other visualizations. Rather, GIS will be used in a conceptually far simpler way as a vital piece of e-research infrastructure.

All humanities research sources contain key elements that can be used by a GIS. First, information has a spatial location whether expressed precisely or generally. Second, all humanities data contains a temporal marker, even though its granularity may vary from minutes to a decade, century, or more. Thus, in most instances in the humanities, research information is placed within a chronological framework. Third, all humanities data can be classified by subject. In fact it is only this last facet of humanities research data that has received much attention to date thanks to the work of library and information management professionals. In the analog world libraries organize monographs and manuscripts by subject area. In effect, the same is done with journals, with each one having a broad, or specific, area of disciplinary concern and collectively grouped around thematic similarities. In the digital world there is the opportunity to organize information in a far more complex way using not only the subject nature of information, but its chronological and spatial attributes as well. GIS is able to treat date and location as an attribute and as such allows for more exacting organization of information than the traditional subject-based approach. This functionality, coupled with the proliferation of electronic resources, offers new opportunities for resource discovery and use by humanities scholars. It will result in a step change in the access humanists have to research information compared to what was ever possible in the analog world. A key toolkit which will bring these seemingly endless and unmanageable arrays of data to the humanist are Grid technologies, and it is to these approaches that we now turn.

E-SCIENCE AND GRID TECHNOLOGIES

It is likely that most humanities scholars have had no dealings with e-science or associated grid technologies. In many ways the term e-science is sufficient to convince humanists that this area of methodological innovation is not of any interest or relevance to them. That, added to the fact that both e-science and the grid have received by far most attention in the sciences, only seems to confirm that view. In practice, while challenges remain, e-science and the grid have the potential to revolutionize scholarship in the humanities.

What is e-science? The United Kingdom Arts and Humanities Research Council defines e-science as an overall approach encompassing grid technologies:

> e-Science ... stands for a specific set of advanced technologies for Internet resource-sharing and collaboration: so-called grid technologies, and technologies integrated with them, for instance for authentication, data-mining and visualization. This has allowed more powerful and innovative research designs in many areas of scientific research, and is capable of transforming the [arts and humanities] as well.[1]

There are three elements to grid technologies within e-science—the access grid, the computational grid, and the data grid. The access grid facilitates online collaboration, including, but not limited to, real-time virtual meetings, interactive simultaneous sharing of data with the ability to manipulate research information, and video communication which supports, for example, high-definition viewing of arts installations and performances. In a newly conceived world of collaborative research between humanities scholars, the access grid aims to enhance cooperative working among humanists, not limited to their immediate colleagues. In practice, the access grid involves significant change in the way humanists work. Typically, they do not work as part of a large team, as is common in the sciences and social sciences. Humanists also do not adopt the scientific tradition of sharing research outputs through multi-authored works. While the access grid has the potential to allow location-neutral collaborations, and this may favor some scholars with specialized disciplinary interests whose immediate colleagues do not work in relevant areas, in reality the access grid will only become of importance when the model for research in the humanities changes. Currently, relatively specialized facilities are required to use the access grid effectively, which for the present restricts its impact in the humanities still further.

The computational grid allows access to, as the name suggests, computationally intensive, remote processing facilities and custom software. At present, most work in the humanities does not require either highly specialized software or significant levels of computational power. It is conceivable in the future that humanists will make limited use of the computational grid, however. Possible applications might include complex comparative analysis of the stylistic approaches and changes adopted by artists or comparisons of narrative styles for large volumes of text.

It is the data grid that is likely to have the most impact on the humanities and is of most concern and relevance from a GIS perspective. In the sciences the data grid is normally associated with technologies and hardware to move very large amounts of data stored in one location to a user in another location. Typically, large data holdings are held in vast data silos and the challenge the data grid resolves is moving these data effectively in a short time. In the humanities, datasets tend not to be large in terms of the amount of storage space required. A medievalist might spend a whole academic career carefully transcribing and extracting information from a survey, for instance the 1377 lay subsidy or crop yield data gathered from multiple sources, and yet stored as text the results would require very little storage space.[2] Equally, a database of millions of statistical values, as is the case for the 32,000,000 data-value Database of Irish Historical Statistics, would easily fit on a low-capacity USB drive.[3] The normative use of the data grid is likely to grow in the humanities as image bases and video repositories are made available online. Such materials require large storage capacities. At present, however, the power and importance of the data grid is its ability to link disparate online datasets. Such isolated, relatively small, specialized, data silos are typical of humanities e-resources. A perusal of projects funded under the AHRC Resource Enhancement Scheme, a UK program, makes this clear.[4]

The challenge, therefore, is to allow the data grid to interlink these rich but isolated seams of research data. In fact, it is a contest that is far greater than that faced in the sciences. As noted, the problem is not moving large amounts of largely uniform data around. In the humanities the challenge for the data grid is manifold. It needs to be able to discover online e-resources. It needs to link to relevant information. It is not sufficient for a tool to refer simply to a home page of a Web site containing relevant information. The link needs to be to the information itself through deep-Web linking. It also needs to cope with unique or customized dissemination systems, inadequate documentation, and partial or non-standard metadata. Beyond discovering such data, a system will need to retrieve that data for the scholar to examine and, if suitable, use. Finally, the data grid needs to be structured in a way that information can be retrieved in terms of relevant elements. It is currently easy to search for information by subject. Even using a basic search engine such as Google, a simple text-based search is likely to retrieve relevant information if the subject is sufficiently specific, although even then the amount of returned information might be vast. What is difficult is to retrieve information by location and by chronology and it is here that humanities GIS, using e-science (henceforth referred to simply as humanities e-science in this specific context), has much to offer, particularly with an implementation that can retrieve information held deeply within a database-driven Web site.

DEVELOPING INFRASTRUCTURE

We are, therefore, currently faced with dealing in a productive way with two new technologies that need to be applied collectively to address a new paradigm. On the one hand, we have humanities GIS, which so far has largely been used by the few as a mapping tool mostly for quantitative data. On the other hand, we have the data grid, which in the humanities is largely unused and in the sciences tends to be employed to move around large amounts of data from a relatively discrete number of sources. However, the ongoing development of client-server protocols such as Z39.50 which offers the opportunity for discrete humanities e-resources to "connect" to each other, irrespective of the server structure, based on client queries.[5] The challenge is to use GIS not as a mapping tool but as a data management tool, although some scholars may choose to visualize their data as well. Many GIS specialists describe a GIS as a three-dimensional database, the third dimension being space or location, but they rarely if ever consider that this functionality might be used without visualizing the results.[6] Tangentially, e-science specialists have given relatively little attention to the opportunity to use the data grid as a resource discovery and retrieval tool to manage the data deluge. Key to realizing the promise of humanities e-science is the development of infrastructure that will assist in this function. Once data are retrieved, they can generally be used within a GIS without many further methodological advances. In most instances, though, scholars will not require further GIS functionality.

In the humanities locations are often designated in colloquial terms by a place-name. These references may be vague or quite precise. They might reference datasets at a national level—information for the United States for example. At the other extreme, they might contain a precise grid reference such as a local ten-figure U.S. National Grid (USNG) reference accurate to a single meter, or a collection of grid references to identify an administrative unit. At its most precise, these grid references may be "date-stamped" to take account of changes in administrative boundaries over time. Most commonly, however, arts and humanities e-resources are referenced by a place-name. Place-names thus form a common language in which to interrelate and associate e-resources. In fact, even when the most precise references to a location, as either a point or polygon are given, these data also are accompanied by a place-name. It might appear straightforward, therefore, to use the data grid and place-names to retrieve relevant e-resources across the humanities. However, this is not the case.

Current barriers to linking sources by place are many. There is no "official" record of place-names in existence. Many national mapping agencies hold a comprehensive gazetteer of place-names but, large as these may be, the resource is of limited value. Crucially, it will contain only modern place-name spellings. Place-names change significantly over time currently making it impossible to use a modern place-name spelling and hope to retrieve relevant e-resources relating to an earlier spelling. Work has been done to create historical gazetteers. In the United States the National Historical Geographical Information System (NHGIS) has, in effect, developed an administrative geography historical gazetteer from 1790 to 2000.[7] Much the same has been done in the United Kingdom, as demonstrated later by examining the Vision of Britain through Time exemplar project. As with the Vision of Britain and the NHGIS, however, almost all historical gazetteer work has concentrated on administrative place-names most commonly associated with the collection of official statistics such as demographic censuses. Such gazetteers may contain tens of thousands of place-name spellings and may record the changes in the spelling of names over time. When linked to a historical geographical information system, they will also likely plot changes in the boundaries and area of the administrative unit as well. Useful as this is, place-names that do not correspond to administrative geographies will not be included. Thus, place-names of hamlets, farmsteads, and the like, which, in the UK for example, were never British administrative units, will not appear in an administrative gazetteer. Nor will topographical names such as rivers, lakes, canals, and so on. Equally administrative geographies typically have little chronological depth. With the exception of a handful of countries, such as China, clearly defined administrative geographies are modern, dating largely from the late eighteenth century or the nineteenth century.

It is clear, therefore, that to utilize the combined power of GIS and e-science, there is a need for place-name infrastructure, but currently that infrastructure does not exist. There are two approaches to resolving this problem. The first is to make existing place-name research available online, in effect to use this information to develop an authority file of names. The second is to extract place-names from existing e-resources and associate them manually and in a semi-automatic fashion using Web 2.0 techniques.

In England a vast amount of place-name gazetteer work has been conducted by the English Place-Names Society (EPNS) based at the University of Nottingham.[8] Since 1924, EPNS has been engaged in the painstaking collection and analysis of all the England's place-names, including the names of administrative units, settlement sites, topographical features, field-names, and street-names (Figure 9.1). In their own right, the place-names are of value and can result in substantive research. Names can, for example, reveal much about the cultural and social patterns of English history: the suffix by, for instance, can be seen to chart the Scandinavian settlements of the ninth, and tenth centuries, while Celtic survival in Anglo-Saxon England is marked by recurrent instances of Walton "the settlement of the Welsh." Topographical names and field-names record changes in landscape and land-use.[9] Not only has EPNS conducted this detailed work for seventy-five years on the origin and derivation of place-names, but critically it has systematically collected place-name spellings from a wide array of textual sources: these spellings are arranged chronologically and related to their modern forms. Shifting spellings of the same name are dated; wholesale replacements of the name are recorded. Thus the material provides a very direct authority file to a significant set of primary sources of interest to humanities scholars. This material in its analog form is printed in 72 volumes to date and more than 18,500 pages, and collectively is highly regarded as the standard reference work. The critical importance and indispensable nature of these volumes can scarcely

Great and Little Horwood

HORWOOD (Great and Little) 94 C 12 and 13 [hɔrud]
Horwudu 792 (c. 1250) BCS 264, 1152-8, 1160-5 NLC
Hereworde 1086 DB (sic)
Horrewde 1227 *Ass*
Horewode 1228 WellsR
Horwode 1301 Ch

These last are the regular forms down to 1502, with the exception of

Harewode 1242 Fees 881, 1461 IpmR
Herewode 1284 FA
Harwod 1509 LP
Horroda (sic) 1512 LP
Herwood 1535 VE
Harwood 1766 J, 1806 Lysons

'Filthy' or 'muddy wood' *v.* horh, wudu. The soil is clayey. The forms with *a*, from the 16th cent. onwards, point to a dialectal development similar to that found in Harpole (Nth), DB *Horpol*, but no trace of it can be found in the present local pronunciation.

GREENWAY FARM

'Possibly takes its name from the family of Greenway who held the manor of Singleborough in the 16th cent.' (*VCH*).

NORBURY COPPICE

The *-bury* is a Camp, 'an almost rectangular work...not shown on the O.S. maps' (HMN 178).

FIGURE 9.1. The English Place-Name Society entry for Great and Little Horwood, Berkshire. A. Mawer and F. M. Stenton, eds., *The Place-Names of Buckinghamshire* (Nottingham: Survey of English Place Names Society, 1996), 68.

be overstated. In digital form this source would provide a powerful tool to integrate disparate resources by place using the data grid. Although so much scholarship, over so long a period, is to be welcomed, the EPNS volumes currently do not exist digitally. It would be a major undertaking to computerize them and extract the data. In any data capture program total accuracy would be essential. In short, there is little point in capturing varying place-name spellings if errors are introduced at the transcription stage. Further, millions of names are included and the volumes, while having a basic standard format, cannot be bulk digitized. While an enormous project, with the development of e-science, it is a worthy one and discussions are being held to begin this work.

England is fortunate in that so much detailed place-name work has been carried out. The same cannot be said for many other countries. The United Kingdom is relatively well provided for in that EPNS has sister, if less developed, projects, in Northern Ireland and Wales.[10] A similar project at EDINA, Edinburgh University in Scotland, uses a basket of sources to identify place-names.[11] Work is taking place in a number of other European counties, but without doubt the EPNS research is most advanced and so in the United Kingdom the development of e-science infrastructure based on detailed scholarly research on place-names taking decades is a sensible route to follow.

Where such information does not exist, one option is to develop a gazetteer from a partial source. The most obvious, and commonly existent sources, are indexes from censuses. Censuses either establish their own reporting geographies or use existing administrative geographies. In many instances they do both, borrowing administrative units for higher-level larger reporting areas while inventing new units for smaller collecting areas. All censuses need to be clear which unit is where and what its name is. Commonly, censuses also nest smaller units in larger ones, thus forming a geographical and administrative hierarchy. They also track changes in both place-names and geographies from census to census to facilitate a commentary on changes between different censuses. This feature leads to time-enabled indexes of reporting units in published census returns. It is important to remember that these are reporting units and not all places (however defined) will be included. Obvious places excluded from census geographies are physical features such as rivers, lakes, other waterways, mountains, hills, colloquially phrased areas, and so on. It is, though, precisely because census geographies are just that, and because they exclude so many other places, that they can be relatively easily gathered. Further, censuses in subsequent years will reflect both the creation of new places—in other words new administrative units for which data are collected. These tend to be in growing urban areas, depopulating rural

ALPHABETICAL INDEX TO THE TOWNLANDS AND TOWNS OF IRELAND.   485

No. of Sheet of the Ordnance Survey Maps.	Townlands and Towns.	Area in Statute Acres. A. R. P.	County.	Barony.	Parish.	Poor Law Union in 1857.	Townland Census of 1851, Part I. Vol.	Page
34, 35	Glenedra	2,150 1 18	Londonderry	Keenaght	Banagher	New T⸗Limavady	III.	234
78, 87	Glencely	292 0 24	Donegal	Raphoe	Donaghmore	Stranorlar	III.	138
44, 45	Glenceny	586 1 19	Tyrone	Omagh East	Termonmaguirk	Omagh	III.	314
6, 10	Gleneige	514 3 6	Leitrim	Drumahaire	Drumlease	Manorhamilton	IV.	94
8, 13	Glenerin	2,014 2 11	Tyrone	Strabane Upper	Bodoney Upper	Gortin	III.	324
71	Glenfad	265 2 10	Donegal	Raphoe	Clonleigh	Strabane	III.	135
47	Glenfield	428 2 39	Limerick	Kilmallock	St. Peter's & St. Paul's	Kilmallock	II.	250
6, 7	Glenfield North	136 1 23	Cork, E.R.	Duhallow	Knocktemple	Kanturk	II.	75
6, 7, 15, 16	Glenfield South	343 3 1	Cork, E.R.	Duhallow	Knocktemple	Kanturk	II.	75
42, 45	Glenfin	43 2 32	Roscommon	Athlone	Rahara	Roscommon	IV.	183
34	Glenfinshinagh	94 3 36	Tipperary, N.R.	Kilnamanagh Upper	Upperchurch	Thurles	II.	280
20	Glenfooran	119 1 20	Waterford	Coshmore & Coshbride	Lismore and Mocollop	Lismore	II.	346
11	Glenga	379 0 31	Tyrone	Strabane Upper	Bodoney Upper	Gortin	III.	324
21	Glengad	105 3 19	Antrim	Kilconway	Finvoy	Ballymoney	III.	26
2, 4, 5	Glengad	2,569 0 33	Donegal	Inishowen East	Culdaff	Inishowen	III.	118
3, 4, 10, 11	Glengad or Dooncarton	830 2 23	Mayo	Erris	Kilcommon	Belmullet	IV.	143
70	Glengaddy	128 0 12	Tipperary, S.R.	Middlethird	Barrettsgrange	Cashel	II.	325
55	Glengall	661 1 28	Tipperary, S.R.	Slievardagh	Ballingarry	Callan	II.	331
44, 45	Glengar	788 1 29	Tipperary, N.R.	Kilnamanagh Upper	Doon	Tipperary	II.	277
74, 80	Glengarra	1,029 2 11	Tipperary, S.R.	Iffa and Offa West	Shanrahan	Clogheen	II.	319

FIGURE 9.2. Irish townlands in their administrative hierarchy.

areas, or through whole-scale changes to administrative geographies, as happened in Britain in 1974.

For the Vision of Britain through Time project, discussed in detail below, places initially were based on census geographies, reflecting the background to the work which created the first national historical GIS. The United States NHGIS followed a similar path. Both include place-names as they changed over time and reflect new units as urban areas expanded. They still do not contain, however, most geographical features. A search for River Thames produces no name associations and neither does Thames, showing that this is not, and never was, an administrative area.

In other countries, census-based gazetteers can be more valuable. Many Irish censuses publish indexes of place-names (Figure 9.2).[12] These tables also contain hierarchal information. Thus, a townland name is recorded and the parish, barony, and county in which it appeared are also noted. This sort of information is exceptionally useful. It offers the possibility of using humanities e-science in a more sophisticated way. A user might be interested in information for a particular townland, but there are 62,000 townlands in Ireland. A gazetteer could easily be implemented based on the Irish census tables that also searched for the parish in which the townland was based. Thus, in any document the name of the townland in question would not need to appear, just the name of the parish. The Irish census geography is exceptionally detailed and this too aids its use as the basis for a place-name gazetteer. Ireland is roughly 80,000km^2 and with 62,000 townlands, 2,000 parishes, 340 baronies, and 32 counties, such a gazetteer results in a large number of names. By comparison in England and Wales there are only 16,000 parishes (the smallest published historical census reporting unit) but the land area is almost double the size at around 150,000km^2. Moreover, there are more places reflecting the difference in population size. In Ireland in 2001 the total population was a little less than 6 million whereas in England and Wales it was 52 million.

So, for some counties more than others censuses can be used as a basis for a gazetteer for humanities e-science grid infrastructure. There is an alternative from this "top-down" approach to gazetteers and that is to let existing e-resources define their own gazetteers. This involves the semantic Web and Web 2.0. Such an approach is best suited to less structured, more complex, and increasingly more common sources, such as text.

Much work on marking up place-names in text is represented by the Text Encoding Initiative (TEI), which forms a consortium of projects and institutions that have developed certain standards in text encoding.[13] Where location names have been flagged, these standards allow textual sources to be mined and the place-names extracted. The application of TEI is a skilled and time-consuming task, however, and, although much work has been done, in terms of results only a relatively small proportion of online texts have been marked up in such a way. There are further issues with TEI for our purposes. Ideally, as noted earlier, humanities e-science requires not just that a place-name be flagged and identified, but also that the magnitude of the place be recorded—its position in a hierarchy, for example, in Ireland a certain townland falls within a certain parish in a certain county. TEI as generally implemented lacks this hierarchical structure in terms of its coding, so a place-name would typically be highlighted as:

XXXXX

thus not reflecting the position of the place in a wider structure of places. TEI has further problems in that different concepts of places are not well represented. A gallery in Paris may contain paintings of Florence, for instance. But we generally are interested in the location of the gallery

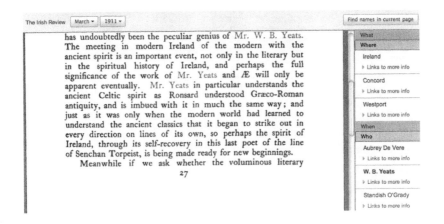

FIGURE 9.3. Person names, semantic searching, and linking information.

PARIS rather than what the painting shows FLORENCE, although we might also be interested in both locational elements. TEI fails to differentiate between these different meanings of place.

As a result, using TEI texts to extract place-names and develop a ground-up gazetteer is unlikely to prove fruitful. The encoding is not sufficiently complex for our needs, and the texts too few. The semantic Web and Web 2.0 approaches appear more promising. To clarify, searching semantically is concerned with using contextualization to identify likely place-names and then flagging them. This is best applied to large corpuses of text. Under a National Endowment for the Humanities (NEH) grant, "Context and Relationships: Ireland and Irish Studies," work is underway at the University of California at Berkeley, with Queen's University Belfast, to develop and apply this technology to place-names and person-names for a large body of Irish Studies Journal material developed at Queen's.[14] The text repository, largely in free text format, is composed of around 600,000 pages of journals drawn from 80 current and historical titles, more than 200 monographs and 2,500 pages of manuscript material.[15] The common factor is that they all relate to Irish studies but, as such, cover a wide range of specific humanities disciplines including English, human geography, sociology, linguistics, politics, religious studies, literature, and philosophy. The work aims to search this vast body of material in a non-traditional way by place and person name. To date, it has proven

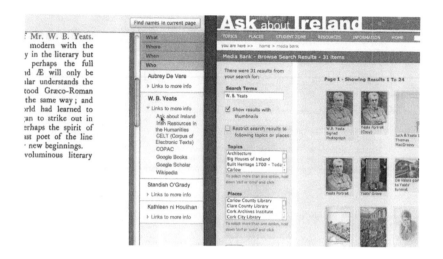

FIGURE 9.4. Confirming the identity of W. B. Yeats.

to be remarkably easy to extract person and place names with high levels of accuracy.

This work achieves what TEI aims to do but in a fraction of the time. It builds on existing computational linguistics and more specifically named entity recognition work to automate the recognition of places and people.[16] However, as with TEI, it results in a large number of probable place names but with no clear idea of where the places are located. Here Web 2.0 has an impact. In essence we are concerned with Web 2.0 in relation to Internet users creating networked content. In a beta interface developed for person names (see Figure 9.3) a person name is identified and users are given the option of selecting links to some established reference works. If a reference proves to be correct the user can mark the original text accordingly.

So in this instance using semantic searching of the journal *Irish Review*, the poet W. B. Yeats is identified. It should be noted that all of the variant versions of the name are highlighted. Toward the right of the diagram are links to a range of reference sources including those for subject (what), where, when (chronology), and who. By following these links a user can confirm whether Mr. Yeats is indeed the poet and code the original document accordingly. A link to the Web site "Ask about Ireland" is shown in Figure 9.4. Exactly the same approach is used for place-names

| Baile an chuilinn] | [75] | [Baile an mhóta |

b. an chuilinn; Fm. iv. 118, Au. iii. 278, 280; "prob. Ballinkillin, in b. Boyle, c. Rosc.," Dr. B. MacCarthy.
 b. an chuilinn; Ballinkillin, in p. Bagnalstown, c. Carlow.
 b. an chuirrigh; in Mun.; four Seisreachs Slige belong to Baile an Chuirrig, Fer. 153.
 b. an chuirrin; land of Sliocht Tomais a Burc, Fa. 2.
 b. an daingin; in Corco Duibne, in W. Mun., Ai. 78 a; now Dingle, in Kerry.
 b. an deonach; a castle on W. bank of L. Measg, Wc. 49.
 b. an deoraidh; Ballindore, nr Applecross, Scotl., Sk. ii. 412.
 b. an doire; Ballinderry, in b. Ballintober S., c. Rosc., Ci.
 b. an doire; Ballinderry, d. Down.
 b. an donnanaigh; in Mun., Fer. 153.
 b. an draoighín; Ballindreene.
 b. an drochid; O'S. III. v. 5; Oppidopons, in Roche's country; Ballindrehid al. Bridgetown, on the Blackwater, in b. Fermoy, c. Cork.
 b. an duibh; Ballioduff, tl. in p. Killcoona, b. Clare, c. Galw., Fm. iv. 1064.

Ballingarry, b. Upper Connelloe, c. Limk., to distinguish it from Ballingarry, in b. Coshlea, c. Limk., Au. iii. 506.
 b. an gharrdha; Fer. 82; B. an Garr(da) contains a half seisreach of land, Fer. 153.
 b. an gharrdha cnuic síthe una; Ballingarry, b. Lower Ormond, ¼ m. from conspicuous hill of Síde Abna, 4 m. E. of Borrisokane, c. Tipp., Ods. 579; in Ormond, Fm. vi. 2094.
 b. an ghlenna; in N. Clann Uilliam, in Connacht, Fir. 805; Ó Radubáin of B. an Ghleanda in Uib Amalgaid, Fir. 273, Lec. 168, Fa. 4; Fy. 164, Ballinglenn, ruined castle in vale of Ballinglen, p. Doonfeeney, b. Tirawley; al. Glenn an chairn, Fy. 220, 478, Pgi. ii. 752; B. in Ghlenna is na Baireadachaib, Tirarhalgaid, Hz. 75 b.
 b. an ghleanna; land of Barún Chaisléin Chonaing, c. Luimnig, Hb. 13 b.
 b. an ghronta; land of Baron of Caisleán Conaing, c. Luimnig, Hb. 13 b.
 b. an ghruagaigh; Ballingrogy, or Ballengruogy, nr Crossmolina in Mayo.
 b. an ghuirtín; belonging to Burke, of Castle Conaing, in c. Limk., Hb. 13 b.

FIGURE 9.5. *Onomasticon goedelicum locorum et tribuum Hiberniae et Scotiae;* an index, with identifications, to the Gaelic names of places and tribes showing place names recorded in various sources.

but the link would be to a place-name gazetteer with a spatial reference attached. In this way the vast amount of text that is becoming available through innovations such as Google Scholar can be spatially enabled and searched. The very process of users identifying and adding dynamic links to a gazetteer enrich the gazetteer itself. The gazetteer, rather than holding place names in isolation, can link directly to sources which contain that place-name. This is very much the digital equivalent of the English Place-Name Society's published volumes detailing sources in which a place name appears. As with the analog EPNS volumes, the gazetteer would be enriched through capture of variant place-name spellings as recorded in the linked sources. The enhanced gazetteer becomes capable, through semantic searching, of identifying still more place names reflecting the variant spelling. Within Web 2.0, in essence, these processes become a virtuous circle. The work could be further enhanced through the digitization of existing analog place-name resources relation to Ireland. A key example is Hogan's *Onomasticon goedelicum locorum et tribuum Hiberniae et Scotiae;* an index, with identifications, to the Gaelic names of places and tribes,

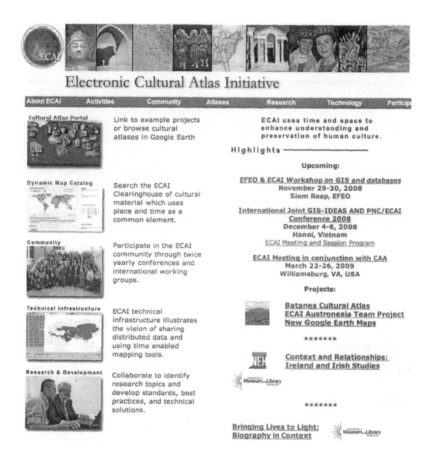

FIGURE 9.6. Electronic Cultural Atlas Initiative homepage.

which, published in 1910, pre-dates the EPNS model in recoding place names as they appeared, in this instance, in 175 sources (Figure 9.5).[17]

It might appear that discussion of the potential for humanities e-science is no more than theoretical, although the Context and Relationships NEH project is concerned with the development of a full proof-of-concept model. Other models already exist, however. The development of custom gazetteers in the Electronic Cultural Atlas Initiative (ECAI) and Vision of Britain through Time projects offer exemplars of how information may be organized over the grid and queried by place and chronology, as well as by the more traditional subject headings.

THE ELECTRONIC CULTURAL ATLAS INITIATIVE

Over the last ten years this innovative initiative has been in the process of implementing e-science in the humanities during a time when e-science was an unknown moniker generally and GIS as a technology was almost completely unused by humanists.

The Electronic Cultural Atlas Initiative is a federation of approximately 1,000 humanists drawn from almost all humanities disciplines and sub-disciplines and representing every continent.[18] ECAI is composed of a number of key elements. First, it has a data clearinghouse which provides interactive, on-the-fly access to a range of spatially referenced humanities datasets provided by scholars from around the world. While having some capacity to store key datasets, the main role of the clearinghouse is to act as a portal to information stored elsewhere. To fulfill this role datasets, or collections of cultural objects, are required to have the relevant ECAI-developed metadata attached to them. While it is possible for the clearinghouse simply to contain collection-level metadata, in order to achieve the level of humanities e-science functionality envisioned in this paper, object-level metadata is required. This is entirely possible, and indeed encouraged, within the ECAI metadata schema.

A second key element to the ECAI suite of resources is a piece of software—TimeMap.[19] Developed by a consortium of scholars based at the University of Sydney, TimeMap is designed to operate seamlessly with the datasets registered in the metadata clearinghouse. In essence it is a temporal geospatial data browser that allows datasets referred to in the clearinghouse to be retrieved and visualized on the fly. In so doing it is able to correctly scale the information, gather it at object level, and display it on a map. It goes further in being able to represent a range of spatially referenced objects at the same time. As noted earlier, this is critical with the ever-growing array of multimedia materials of interest to humanists. Thus, statistical data may be represented, as may text, video footage, and static images. TimeMap, as the name of the software suggests, also allows information to be viewed within a chronological framework. A scholar using the software may choose to retrieve and view material for a particular subject, a particular spatial location, and a particular time.

It might be argued that the ECAI model is the solution to humanities e-science and that it has been in existence for many years. The work of ECAI has been visionary, but regrettably it does not fulfill the potential of e-science. First, the ECAI metadata clearinghouse does not, in some serendipitous way, discover relevant humanities e-resources over the Internet. It will only provide access to e-resources which have been registered. In effect, registration requires the application of metadata to the resource. As noted above, what is needed is deep linking to a resource so that clearinghouse retrieves Romano-British pottery shards from a particular location, and perhaps of a specific type and time period, for example, rather than merely identifying Web sites with shards of this type. The metadata is quite able to do this but the metadata has to be applied first. Since the ECAI metadata does not on the whole follow any of the several metadata standards, it is likely that this detailed object-level metadata will need to be newly applied. This is an onerous task which would take many hours, days, or even weeks. As a result the clearinghouse contains very few, on a global scale, datasets. Thus a scholar could not use the clearinghouse and expect to find every online resource relating to Romano-British pottery. Far more likely, the clearinghouse will not contain any such datasets when many exist. A simple Google search will reveal some of them.

Second, TimeMap is an excellent geospatial data browser but it is limited by the information contained in the clearinghouse. Simply, if there are no data then TimeMap is of no value. That said, it is quite possible for the user to input their own data into TimeMap, but this is a somewhat time-consuming process. Although TimeMap has significant functionality it does not have anything approaching the functions a full GIS software package such as ArcGIS might have. TimeMap is free and open source whereas ArcGIS is expensive and propriety software, but because it is free and largely maintained by interested scholars without payment, its ongoing existence is questionable. As a result of its lack of support TimeMap can be unstable and has limited documentation. Updates are sporadic at best. Any scholar wishing to expend effort in entering data into TimeMap when the clearinghouse cannot provide relevant data would be better using proprietary GIS software. Further, such scholars are likely to be sufficiently interested in using GIS to justify the time taken to master relevant elements of software such as ArcGIS. The real strength of TimeMap is as a tool to link data by location and visualize it in a limited way. It will not produce good maps—indeed it is difficult to print much more than a screen dump of the display on the computer monitor.

So, at face value, ECAI has a humanities GIS e-science solution. Through the clearinghouse disparate data can be retrieved, associated by location, and visualized if the scholar so desires. The lack of data in the clearinghouse, and the lack of longevity of TimeMap tied to its close association with the clearinghouse, limits its value. The situation would radically change if ECAI metadata were widely applied but this is extremely unlikely to happen.

VISION OF BRITAIN THROUGH TIME

Another project which appears to meet the goal of humanities e-science is the Vision of Britain through Time project.[20] This work, developed by the University of Portsmouth with support from Leeds University and Queen's University Belfast, fits a number of our requirements. It manages material by location, chronology, and subject. It contains a range of multimedia material of interest to humanists such as historical gazetteers describing places, historical and relatively modern maps, travelers' diaries, and a range of statistical data. It has the capability to include photographs and video footage. While the site does include limited mapping functionality for the whole of Britain, in essence it is concerned with retrieving information by location. Many users will never produce a map online from this information. Indeed, such would be the stress on the site server, such functionality does not exist.

Vision of Britain is an excellent exemplar of how humanities e-science could work. However, just as with the Electronic Cultural Atlas Initiative, it has some limitations related to its extensibility. It does not make use of the data grid and the possibility of serendipitous discovery of e-resources for a particular location. It does contain some information for every administrative unit in Britain, impressive in itself, but it is a closed world. The site does not contain any information that has not been put into it by the Vision of Britain team. In practice material has been added manually, or in batch mode, and interlinked manually with different electronic materials. The core of the Vision of Britain is a place-name gazetteer which was originally developed as part of the Great Britain Historical GIS.[21] Sources are

FIGURE 9.7. The Vision of Britain through Time homepage.

semi-manually linked to this gazetteer. So, although information is available for almost every place, that information is limited to what has been included by the Vision of Britain team. Again, it affords an insight into what in grid environment GIS e-science might offer, but it is illusory.

TOWARD GIS E-SCIENCE

We have seen that an ever-increasing range of information is being made available for scholars in the humanities. Already there exists a plethora of digital resources either specifically for humanities scholars or of research interest to them. The only factor that most of these resources have in common is that they were designed and developed in relative isolation and that very few are interoperable or cross-searchable. The traditional approach has been to do nothing to encourage interoperability. The United Kingdom Arts and Humanities Research Council Resource Enhancement Scheme, for instance, allowed scholars to develop silos of very specific research data with little interest in encouraging its use with other sources. These resources were developed at not inconsiderable cost with funding up to $500,000 for individual projects. More recently Information Management (IM) practitioners have encouraged the adoption of metadata standards at collection level. As we have seen, object-level linkage is required and, more recently still, IM practitioners have supported this approach. In truth the application of such detailed metadata is a very significant task and, as a result, it is not widely applied. The TEI community, with dogged determination, has encoded texts, often in remarkable detail. We have seen, however, that its treatment of place-names does not suit the needs of humanities e-science. Moreover, despite their efforts only a small proportion of e-content of importance to the humanities has received the attention of the TEI community.

The future does not look promising at first sight. In the past e-resources tended to be text-based, and most often numeric, drawn from key sources such as censuses. The very nature of this information, which in analog format had a clear structure, resulted in much the same in the digital world. Some work was even done to make such material more useful by the adoption of methodologies only possible on any scale digitally. Thus statistical data from one census could be compared with the same information for a different census. In many ways the high-point of historical GIS was the ability to redistrict data between different censuses and different administrative units through a variety of techniques. We now have many more text resources, most of them alphanumeric and in a generally unstructured format. The work of Google Scholar reflects this. What structural information, or metadata, that is available is mostly at collection level rather than object level. The availability of video footage and audio recordings pose still more challenges. Yet more significant is the step-change in the way e-resources have been created. In the past large collections of research data were developed by technically savvy scholars or IM professionals. These resources tended to have some structure and some metadata. Today most humanists are creating e-content—from a journal article made available by JSTOR to small datasets relating to their research. The rich and varied nature of these data is radically different from the situation just ten years ago.

There are, as this essay has noted, significant opportunities however. Information infrastructure is becoming more transparent to facilitate the linkage from a technical perspective of disparate e-resources. The development of e-science and particularly the data grid make this apparent. Still further, the adoption of open standards to facilitate interoperability, such as Z39.50, reduce the technical barriers. Of greater importance still is a changing mindset amongst humanists. Whereas in the past a scholar may have carefully guarded their research data reserving it for their own use, increasingly scholars see the benefits of sharing information, both in terms of receiving recognition for their scholarship and through advances in their own work using information beyond their ability to collect. In teaching, in the United Kingdom, Jorum is a clear mark of this.[22] Jorum aims to develop a community of teaching scholars to share and develop learning and teaching materials. Humanists are prominent among the more than 400 participating Higher and Further Education institutions.

The technical and mindset barriers are thus diminishing while electronic information is exponentially increasing. There has never been a better time to take advantage of these fortuitous happenstances. There is an unprecedented opportunity for humanities GIS to become mainstream in managing data in the developing information environment using all three characteristics of research data—location, chronology, and subject. The exemplars described in this paper—the Electronic Cultural Atlas Initiative and the Vision of Britain through Time—show the potential of humanities GIS. The route forward is demonstrated by the UC Berkeley and Queen's Belfast Context and Relationships NEH-funded project, and through the development of authority files. We have seen several approaches to the development of such an infrastructure that range from the development of gazetteers from detailed scholarly analog resources to the use of Web 2.0 methodologies.

Humanities scholars are not generally known for their eager adoption of new technologies and research methods. There are, of course, innovators and many of the essays in this book represent the work of such scholars. As noted in the introduction to this essay, however, GIS will never become an indispensible technology if it is concerned primarily with information visualization. But conceived as a tool to manage information, it will become an important humanities tool. Its value, moreover, will be enhanced because it will require scholars to do nothing new. They will not need to change their research practices. GIS e-science will simply deliver more and better information, oriented to subject, time, and location, to improve their research. Exciting times are upon us.

ACKNOWLEDGMENTS

This paper represents the work of many and although the author has actively collaborated with most, if not all, of the projects cited, he has by no means taken the lead. Acknowledgment is due to colleagues at the Electronic Cultural Atlas Initiative and particularly to Professor Michael Buckland and Professor Lewis Lancaster. For the section discussing TimeMap the tireless work of its chief proponent, Dr. Ian Johnson of the University of Sydney, needs to be recognized. Dr. Humphrey Southall of the University of Portsmouth has invested vast amounts of time and effort to develop the Vision of Britain project. The opportunity to consider some of the possibilities afforded by the work of the English Place-Names Society at the University of Nottingham has been invaluable.

NOTES

1. Arts and Humanities Research Council, *e-Science:* http://www.ahrcict.rdg.ac.uk/activities/e-science/background.htm (accessed Dec. 2008).

2. *Three Centuries of English Crops Yields 1211–1491:* http://www.cropyields.ac.uk/ (accessed Nov. 2008).

3. History Data Service, *Census Statistics:* http://hds.essex.ac.uk/history/data/census-statistics.asp (accessed Dec. 2008).

4. Arts and Humanities Research Council, *Resource Enhancement Scheme*: http://www.ahrc.ac.uk/FundedResearch/Pages/ResourceEnhancementSchemeReview.aspx (accessed Nov. 2008).

5. *Z39.50 Maintenance Agency Page:* http://www.loc.gov/z3950/agency/ (accessed Dec. 2008).

6. Ian N. Gregory, and Paul S. Ell, *Historical GIS: Technologies, Methodologies and Scholarship* (Cambridge: Cambridge University Press, 2008), 21–40.

7. Minnesota Population Center, *National Historical Geographic Information System: Pre-release Version 0.1* (Minneapolis: University of Minnesota, 2004).

8. Institute for Name-Studies, *The Survey of English Place-Names:* http://www.nottingham.ac.uk/english/ins/survey/ (accessed Dec. 2008).

9. O.J. Padel, and David N. Parsons eds, *A Commodity of Good Names: Essays in Honour of Margaret Gelling* (Nottingham: Shaun Tyas, 2008).

10. Queen's University, Belfast, *Northern Ireland Place-Name Project:* http://www.qub.ac.uk/schools/SchoolofLanguagesLiteraturesandPerformingArts/SubjectAreas/IrishandCelticStudies/Research/NorthernIrelandPlace-NameProject/ (accessed Dec. 2008).

11. EDINA, *Geographical Information*: http://www.geo.ed.ac.uk/ (accessed Oct. 2008).

12. *Census of Ireland for the Year 1861: General Alphabetical Index for the Townlands and Towns, Parishes, and Baronies of Ireland* (Dublin, 1861).

13. TEI: Text Encoding Initiative: http://www.tei-c.org/index.xml (accessed Dec. 2008).

14. Electronic Cultural Atlas Initiative, *Context and Relationships: Ireland and Irish Studies:* http://ecai.org/neh2007/index.html (accessed Nov. 2008); National Endowment for the Humanities, *Advancing Knowledge: The IMLS/NEH Digital Partnership Grants:* http://www.neh.gov/news/archive/20070917.html (accessed Dec. 2008).

15. JSTOR: http://www.jstor.org (accessed Dec 2008); JSTOR, *The Ireland Collection:* http://www.jstor.org/templates/jsp/_jstor/templates/info/about/archives/irelandHandout.pdf (accessed Nov. 2008).

16. CoNLL-2009, *Thirteenth Conference on Computational Natural Language Learning:* http://www.cnts.ua.ac.be/conll/ (accessed Jan. 2009).

17. Edmund Hogan, *Onomasticon goedelicum locorum et tribuum Hiberniae et Scotiae; An Index, with Identifications, to the Gaelic Names of Places and Tribes* (Dublin: Hodges Figgis, 1910).

18. The Electronic Cultural Atlas Initiative: http://ecai.org (accessed Dec. 2008).

19. TimeMap Open Source Consortium, *TimeMap: Time-Based Interactive Mapping:* http://www.timemap.net/ (accessed Dec 2008).

20. A Vision of Britain through Time: http://www.visionofbritain.org.uk (accessed Dec. 2008).

21. Humphrey Southall, "A Vision of Britain through Time: On-line Access to Statistical Heritage," *Significance*, 4:2 (2007), 67–70.

22. Jorum, *What is Jorum?* http://www.jorum.ac.uk/ (accessed Dec. 2008).

TEN

Challenges for the Spatial Humanities: Toward a Research Agenda

TREVOR M. HARRIS, JOHN CORRIGAN,
AND DAVID J. BODENHAMER

This book set out with the ambitious goal of critically engaging domain experts in the task of examining the role of GIS technology and spatial concepts in the emerging field of the spatial humanities. The resulting chapters explore both the potential of GIS as a core component of the spatial turn and the role of geographical space as a conceptual framework in the humanities. The basic premise under examination is whether the powerful spatial data management, functionality, geovisualization, and mapping capabilities of GIS, combined with a spatial perspective, can provide new insight in humanities scholarship. We have chosen not to develop case studies to illustrate the use of GIS in specific humanities disciplines but rather have sought to identify the present status, history, and nature of the challenges facing the spatial humanities. We also have set out to explore critically the theories, concepts, and methodologies redolent in GIS usage and the spatial analytic perspective in humanities scholarship. In common with humanities scholarship in general, we sought not to develop an authoritative or ultimate answer to the role of GIS in these disciplines but to probe for new questions, develop new perspectives, advance new arguments and interpretations, and recursively shape the interface between GIS and the humanities and between humanists and GIS scientists. In exploring these multifaceted relationships we see the beginnings of a research agenda that is formulated on understanding the core elements of a reoriented humanities scholarship where space and geographical concepts play a greater role in framing scholarship in both the humanities and GIS communities. Likewise, we see opportunities for GIS to experiment methodologically and technically in ways that bring it into closer collaboration with the agendas of the humanities.

The content and themes addressed in the preceding chapters represent the diversity and challenges facing the spatial humanities in all their conceptual and methodological forms. The evident complexity of the spatial humanities and the associated interwoven challenges represent a compelling fascination for this rich field of enquiry. In most instances, grappling with the spatial humanities entails moving beyond traditional disciplinary and methodological boundaries and scholarly comfort zones, and yet it is within this interdisciplinary nexus that the true potential of the spatial humanities becomes most apparent. At the same time as providing a synthesis of issues facing the spatial humanities, the preceding chapters also provide the basis for an evolving and extensive research agenda. Significantly, they discuss both conceptual as well as methodological issues that address, if not redress, the current heavy emphasis given to project and application-driven uses of GIS in the humanities and provide an important and much needed theoretical backdrop to a regrettable over-reliance on technique and method. In so doing, we have sought to give shape and substance to several important aspects of the spatial humanities and to raise a multiplicity of themes that are contextually nuanced and not easily given to summary points. For this reason we conclude with a selection of six themes that mark the nascent field of spatial humanities.

First, latent tension, if not direct conflict, exists in linking a positivist technology with predominantly humanist traditions. On the surface, a humanities GIS could be viewed by many as antithetical, an oxymoron, in that the positivist underpinnings of GIS (though not of geography in its entirety) are misaligned with the epistemological and ontological understandings of the world as practiced by humanists. It could be argued that the GIS and humanities tracks run not in parallel with many switches that allow scholars numerous crossing points to develop linkages between the technology and humanities disciplines, but exist in orthogonal worlds where scholarship is practiced in very different ways and intersect in only in a very few instances that have perhaps already been identified and exploited. Certainly, the early adoption and acceptance of GIS in the sciences and social sciences reflect the compatibility of this positivist technology with these fields of enquiry. The much later entry of GIS into the humanities reflects deeper ontological and epistemological concerns that go beyond mere claims of technophobia and reluctance to embrace the digital world, for there are many aspects of GIS that sit uncomfortably with the ways in which humanities researchers frame investigations and report them in prose.

The juxtaposition of contrasting epistemologies is perhaps most apparent in the characteristics and tools of GIS with its emphasis on the quantitative collection of data, the accurate measurement of geographic location, the importance of coordinate systems, datums and projections, the management and cartographic display of geographical information, and the spatial analysis and modeling of phenomena to identify optimal solution space. The conceptualization and categorization of geographic complexity into entities, fields, objects, attributes, spatial primitives, geometric topology, and database schemas is in stark contrast to the traditions of the humanistic weaving of complex construction, of nuanced emphasis, and of pluralistic methods. The scientific method that underpins GIS with its computational demands for accuracy and precision, a Euclidian coordinate space, and its emphasis on generalization and reductionism contrasts markedly with the humanist emphasis on the individual and the unique, on imprecision, uncertainty, and ambiguity, on the emergent questioning of a complex weave of text, on narrative and method, and on the interlacing textures of experience, memory, and artifact. This fundamental clash of scholastic cultures is not new to either knowledge paradigm and yet has obvious and far ranging implications for any discussion of a GIS-inspired spatial humanities. It is too simplistic to speak solely in terms of a qualitative-oriented humanism and a quantitative GIS, yet the contrasting emphases on conceptual mapping rather than cartographic mapping represents significant challenges to the development of a humanistic GIS. GIS privileges disambiguation in its organization of knowledge whereas the humanities treats knowledge as multivalent, equivocal, and protean.

A second theme concerns the epistemological and ontological implications arising from a GIS informed spatial humanities. The issues facing the spatial humanities bear very close resemblance to the concerns and themes raised by the early social theoretic critique of GIS by geographers in the GIS and Society discourse and subsequently within the sub-fields of Critical GIS and Participatory GIS. Understanding and appreciating the ontological issues that fashion and shape the differing conceptualizations of reality represented by the GIS community and humanities disciplines is central to addressing the latent tensions mentioned above. GIS is embedded in a positivist epistemology premised on an objective reality that can be discovered through a process of observation, testing, and the application of the scientific method. Incomplete data, silences in the data, and structural knowledge distortion pose significant challenges to the users of GIS in the humanities particularly because of the spatial determinant in GIS. GIS imposes a conceptual and logical model of reality that is filtered by the user and by the technology.

What scholars in the spatial humanities should be focusing on is not just GIS as a system, or as a method and technique, but the broader ontological and epistemological issues of GIS as science (GISci) and about how spatial information science brings about a rethinking and additional insight into humanities scholarship. The interface and juncture of GISci with the humanities generates a deeper and significantly more challenging and intellectually rewarding contemplation of the conceptualization and representation of space than does an emphasis on GIS as a spatial tool box. This shift in focus from GIS as system to GIS as science entails a closer alliance in the spatial humanities not just with GIS practitioners but with the discipline of geography as a whole. Focusing solely on the nuts and bolts of space in a GIS, with little understanding of the geographical concepts that lie behind the technology, is like focusing solely on historical dates to the detriment of any deeper understanding of human behavior and events that historical investigation provides. GIS is not a panacea for the humanities; its appropriate use demands good judgment and a broader knowledge of the production of space than the application of a technology can provide. The spatial turn in the humanities must be more than method and there is a need to understand the role of space in human events. To that end the spatial humanities needs to embrace more than geospatial technologies but also the broader discipline of geography and of geographical concepts of space.

Thirdly, GIS is a seductive technology with its reductionist allure and wondrous images. The role of maps and of spatial representation is an important theme for humanists to comprehend. Text is ill-suited to represent the higher dimensionality of spatial information. Maps are a powerful medium to store vast quantities of information and yet enable patterns and relationships to emerge. As geographers have been aware for a long time, map representations are beguiling, subtle, and yet powerful instruments of persuasion that are capable of misleading interpretation. The spatial turn in the humanities is predominantly a GIS-enabled rediscovery of the power of the map, and yet GIS mapping has long been recognized as adept at making bad data look good. Many would, and do, question the words of an author or the verisimilitude of a virtual representation, yet few question the content, symbology, and representation of a seemingly objective map. Cartographic representation is a skilled art and many GIS practitioners perform the duties with little or no understanding of the cartographers' skills and are largely ignorant of the assumptions and impositions that maps impose on ways of knowing. Traditionally in the spatial humanities 2-D maps have been the mainstay of GIS display and product, though increasingly 2.5-D representations with flight simulation capability are now common. The related field of geovisualization ranges from mapping to dynamic and animated maps, Virtual Reality, and cyber geography and has much to offer the spatial humanities as a way of handling vast quantities of complex spatial data and enabling insightful new ways of seeing and understanding spatial relationships. The potential for the integration of qualitative data in the form of images is substantial and currently unexploited. Virtual GIS combines the elements of Virtual Reality and serious gaming, with the spatial analytic and data handling capability of GIS to provide an immersive and experiential environment that is closer to the phenomenological approach favored by many humanists. Reducing the subject-object paradigm embedded in traditional mapping and providing more powerful interactive and navigation tools through innovative geovisualization has much to offer the spatial humanities. It is incumbent on scholars to look beyond the strict confines of GIS and map making per se, important though they be, and discover the other potentially valuable geospatial technologies and techniques that are available to them.

A fourth and related theme is that there are more methods and approaches available to scholars to explore the spatial humanities than the very heavy emphasis on off-the-shelf software packages provided by GIS vendors. The potential role of geovisualization has already been mentioned, but the Geospatial Web as a potential neogeography holds considerable potential for humanities GIS. Not only do Web mash-ups provide an easy and accessible means to generate maps but the Geospatial Web closes the critical gap that currently exists in the humanities between data producer and data consumer. The digital revolution led to a growing assemblage of digital datasets available to researchers. These infrastructural data investments are valuable but cannot hope to be extensive enough to manage the immense diverse data repositories of multimedia materials that are the foundation of humanities scholarship. A GIS based on an integrative Geospatial Web may be better suited to the needs of the spatial humanities than a fully-fledged GIS with its sophisticated yet mostly unused (by humanities' scholars) functionality.

As the capability of the Geospatial Semantic Web unfolds, its potential for the spatial humanities to automate the identification and mapping of people, events, places, and spatial relationships from textual resources is substantial. Similarly, the ability to transform unstructured text into structured maps through computational text and data mining, semantic synthesis, geoparsing, place name matching using natural language processing and digital gazetteers, and georeferencing methods promises to be a major contribution to the armory of tools available to the spatial humanities. Text is the major record of human experience and the conversion of experience into discourse and the space of acculturation bridges the literary and geospatial scientific worlds. Semantic space rather than geographical space becomes important and connects experience with the revelation of reality through spatial story telling. Just as narrative-based story telling dominates the humanities, so spatial story telling can complement one of the major traditions in the humanities. Spatialization techniques such as self-organizing maps and text clouds identify clusters in text documents that share similar characteristics in metaphorical space and not just geographical space. Text map transformations can reflect both absolute space based on Euclidian coordinate systems as well as the relative space of textual association based on place names, spatial rules, and geospatial markers that extract spatial relationships embedded in text and go beyond the strict cartographic map making that dominates current humanities use of GIS.

These developments and possibilities can greatly enhance the machine readable e-resources. While much resource infrastructure development has focused predominantly on primary quantitative and statistical sources, there are nonetheless initiatives underway to address text-based, image, and qualitative sources as well. The value of the Internet and of e-science in this endeavor cannot be underestimated. From resource sharing of online data grids, to grid technologies to facilitate access and collaboration, to computational remote processing facilities, the evolving e-science framework is remarkable for its utility to non-expert users in the humanities. Though considerable interoperability issues surround these e-grid initiatives, e-science reinforces the infrastructure so important to the humanities and provides a major resource with which to undertake space-enabled discovery.

The fifth theme relates to the centrality of time to most humanities disciplines, for change occurs simultaneously through both time and space. The spatial and temporal turns go together, and it is unwise to separate the two or prioritize one over the other. What is clear is that the spatial humanities must push the GIS envelope to develop spatio-temporal tools. While there is need for a spatial framework for humanities research, so too is there need for a combined spatio-temporal capability. One would have thought that for historical GIS the spatio-temporal perspective would be a significant distinguishing feature and yet, because GIS has struggled to adequately handle the complexities of these spatio-temporal needs, time is treated invariably in GIS as categorical and discontinuous.

While the thread of many spatial humanities discussions oscillates between the priorities of space dependency or time dependency, the spatial turn may allow historians to think about time in new ways. The need for a spatio-temporal GIS to explore changes in space through time is apparent. To speak of history as dealing with time, and geography with space, is too simplistic a divide for modern use. Doreen Massey's suggestion that geography is concerned with exploring multiple trajectories through space and time to allow complex stories of how places change is a powerful one, and yet GIS struggles to provide an integrated space-time analytical environment. The value of animated maps has been acknowledged in helping to understand movement as a basic characteristic of human existence but, as with text mining, the humanities pose significant challenges to the GIS community to develop components linked specifically to their needs.

Ultimately, aspects of place are equally, if not more, important to the humanities than the GIS focus on space and spatial analysis. It is the contested constructedness of place—the linking of locale, events, and process that make up local place—and that of place-making and of the cultural practices employed for place-making, which intrigues humanists rather than the perhaps less subtle GIS emphasis on generalized space. Although the humanities have embraced place as part of the spatial turn, humanists have struggled with the analytical technology enabled through the rise of GISci. Equally, the GIS community has struggled with how to handle humanistic elements in the highly structured database schemas. The technology cannot speak to the contingent nature of cultural processes or to the agents of change and transformation and the humanities penchant for dismembering generalizations, rethinking, and recombining. GIS has difficulty managing deep contingency and thick description or the notion that all social life is contingent, implicated, and unpredictable. The domains of the public, private, economic, social, political, religious, and civil military are deeply connected through a deep contingency that fuses place and time and through which structures are articulated to other structures in a cascading, spiraling, or rupture of local social processes in response to structural transformations of power at other scales. Places defined through fine grain studies are important contributions to larger scale processes because of these highly interconnected systems. Mapping is one way to see deep contingency and the rippling and sweeping of specific place-based events across time and space. The focus here is less on the factors of causation than on interpreting the consequences and the resonances of events and as evidenced through the intersection of documented nodes of place, time, and action—where geocoded singular events and larger patterns intersect across time and space, a collage of moments. This deeply layered interpretive history draws heavily on space and place as an organizing framework to understand the world and yet challenges GIS as to how this might be achieved.

How to grapple with deep contingency and with the scaled layering of intricate linkages through space, place, and time is a primary challenge for the spatial humanities and its use of geospatial technologies. Deep mapping might provide one such vision for the spatial humanities in that it represents an intensive topographical exploration of place that weaves a complex multi-layered deep map of both the visible and invisible aspects of a place. Deep maps are heavily narrative-based and interlace autobiography, art, folklore, stories, and memory with the physical form of a place to "record and represent the grain and patina of place through juxtapositions and interpenetrations of the historical and the contemporary, the political and the poetic, the discursive and the sensual. . . ."[1]

One of the best exponents of deep mapping is William Least Heat-Moon who wrote *PrairyErth (a deep map)*.[2] The idea of deep mapping has a counterpart in geography in the work of Yi Fu Tuan's *Topophilia: A Study of Environmental Perception, Attitudes and Values*[3] who proposed exploring the connectedness and ties between human emotion and the physical fabric of landscape and which gave rise to major advances in the study of identity and sense of place in geography. These concepts of deep contingency and deep mapping go beyond the traditional backbone of GIS mapping and point to new realms for pursuing phenomenology in GIS and representing emotion and experience. Barry Lopez situates well the differences between traditional mapping and deep mapping when he writes,

> I would like to tell you how to get there so that you may see all this for yourself. But first a warning: you may already have come across a set of detailed instructions, a map with every bush and stone clearly marked, the meandering courses of dry rivers and other geographical features noted, with dotted lines put down to represent the very faintest of trails. Perhaps there were also warnings printed in tiny red letters along the margin, about the lack of water, the strength of the wind and the swiftness of the rattlesnakes. Your confidence in these finely etched maps is understandable, for at first glance they may seem excellent, the best a man is capable of; but your confidence is misplaced. Throw them out. They are the wrong sort of map. They are too thin. They are not the sort of map that can be followed by a man who knows what he is doing. The coyote, even the crow, would regard them with suspicion.[4]

Humanistic mapping and value-laden mapping runs counter to traditional GIS mapping yet seeks to capture the essence of place and a humanistic sense of distance, direction, and identity. In so doing, deep mapping moves the user from the GIS world of observation to one of habitation where the material world is experienced through our own embodiment and sense of "being in the world." Non-representational theory and the concepts of deep contingency, deep mapping, taskscapes and affordances, and thick description enables scholars to engage the material world rather than observe it, and interrelates theories of practice and agency and how people both create their material world and, in turn, are created by it.

This linking of critical geographies, postmodern humanities, and GISci creates a fresh conceptualization of a humanities GIS. A humanities GIS is a system that is multimedia capable, multilayered, non-authoritative, non-objective, negotiated between experts and contributors, framed as a conversation and not as a statement, is inherently unstable, and is continually unfolding and changing in response to new data, new perspectives, and new insights. Deep mapping is visual, experiential, uncertain, ambiguous, and imprecise and yet lends itself well to spatial multimedia, virtual environments, geovisualization, and potentially GIS. It is a way to pursue a reflexive epistemology that is capable of integrating multiple voices and views and include place-based mini-narratives of small events that are conditioned by unique experiences and local cultures.

We need a broader integration of not just GIS into the spatial humanities but of geography and geographical concepts as well. We envision a spatial humanities that draws on a GIS-enabled fusion of qualitative and quantitative data, that focuses on both space and place, that acknowledges a time embedded space, that is recursive and responsive to exploring multiple layers of dynamic relationships through emergent semantics, that handles nuanced data categories and blends data with judgments, that enables the complex layering and deep mapping of place, that is grounded in human subjectivity as well as objective space, and provides an ambiguated representation of culture with all its contradictions and complexities. What we seek above all is a humanities GIS that is sensitive to the needs of humanities scholars.

NOTES

1. Michael Pearson and Michael Shanks, *Theatre/Archaeology: Disciplinary Dialogues* (New York: Routledge, 2001), 64–65.

2. William Least Heat-Moon, *PrairyErth (a deep map)* (Boston: Houghton Mifflin Company, 1992).

3. Yi Fu Tuan, *Topophilia: A Study of Environmental Perception, Attitudes and Values* (New York: Columbia University Press, 1974).

4. Barry Holstun Lopez, *Desert Notes: Reflections in the Eye of a Raven* (New York: Picador, 1990).

Suggestions for Further Reading

TURNING TOWARD PLACE, SPACE, AND TIME

The streams of writing on space and place are like tributaries that run roughly parallel, flowing from the same intellectual watersheds but seldom crossing or feeding into the other. Any point of convergence seems to lie beyond the borders of our current map.

One tributary is that of critical GIS. The technical and practical literature on GIS is vast, but humanists will be particularly interested in the self-aware and theoretical writing that has already become a rich tradition. Useful guides are Daniel Z. Sui, "GIS, Cartography, and the 'Third Culture': Geographic Imaginations in the Computer Age," *The Professional Geographer* 56:1 (2004): 62–72; Stanley D. Brunn, "The New Worlds of Electronic Geography," *GeoTrópico* (online), 1:1 (2003): 11–29; and Richard Biernacki and Jennifer Jordon, "The Place of Space in the Study of the Social," in Patrick Joyce, ed., *The Social in Question: New Bearings in History and the Social Sciences* (New York: Routledge, 2002).

Another tributary flows from humanistic thinking about space and place. A convenient overview appears in Karen Halttunen's presidential address to the American Studies Association, an interdisciplinary body open to thinking about such issues. See "Groundwork: American Studies in Place," *American Quarterly* 58 (March 2006): 1–15. An elegant, thoughtful, and more general perspective is Denis Cosgrove, *Mappings* (London: Reaktion Books, 1999). Any humanist looking to understand what geography might mean for broader understanding of their field could do no better than the work of D. W. Meinig. His three-volume series on Continental America, *The Shaping of America: A Geographical Perspective on 500*

Years of History (New Haven, Conn.: Yale University Press, 1986–2000) is a masterpiece and his reflections in Douglas Greenberg and Stanley N. Katz, eds., *The Life of Learning* (New York: Oxford University Press, 1994) are inspiring.

Since my essay is about time as well as space, I have followed another stream of theory, on temporal concerns. Andrew Abbott's *Time Matters: On Theory and Method* (Chicago: University of Chicago Press, 2001) is very helpful, as is Donna Peuquet, *Representations of Space and Time* (New York: The Guilford Press, 2002).

One stream not usually in surveys of space and place is nevertheless an intriguing one, practice theory. Here, two books give excellent overviews: William H. Sewell, Jr., *Logics of History: Social Theory and Social Transformation* (Chicago: University of Chicago Press, 2005) and Gabrielle M. Spiegel, ed., *Practicing History: New Directions in Historical Writing after the Linguistic Turn* (New York: Routledge, 2005).

THE POTENTIAL OF SPATIAL HUMANITIES

Compared to the sparseness of the literature on the emergent spatial humanities or the application of Geographic Information Systems to humanities disciplines, works on GIS and GIScience are vast. Readers interested in the history of Geographic Information Systems should consult T. Foresman, ed., *The History of Geographic Information Systems: Perspectives from the Pioneers* (Upper Saddle River, N.J.: Prentice Hall, 1998). N. Schuurman, *GIS: A Short Introduction* (London: Wiley-Blackwell, 2004) is a solid introduction to the epistemological underpinnings of GIS, including Critical GIS theory. An important argument in the critique of GIS may be found in J. Pickles, ed., *Ground Truth: The Social Implications of Geographical Information Systems* (New York: The Guilford Press, 1994).

Humanists will be especially interested in the rich literature on space and place. Geographer Yi-Fu Tuan devoted a career to discussing the connection between these two concepts, as in his leading text, *Space and Place: The Perspective of Experience* (Minneapolis: University of Minnesota Press, 1977). More recent contributions include D. Massey, *For Space* (London: Sage Publications, 2005) and D. Peuquet, *Representations of Space and Time* (New York: The Guilford Press, 2002). T. Cresswell, *Place:*

A Short Introduction (London: Wiley-Blackwell, 2004), provides a useful guide to how scholars have conceptualized space, while B. Janz has compiled a wealth of resources on space and place at http://pegasus.cc.ucf.edu/~janzb/place/ (accessed 26 Jan. 2009). Scholars in cultural studies, film studies, anthropology, and other humanities disciplines discuss how spatial thinking is reorienting their fields in B. Warf and S. Arias, eds., *The Spatial Turn: Interdisciplinary Perspectives* (New York: Routledge, 2008).

For the historical GIS, readers will want to consult I. Gregory and P. Ell, *Historical GIS: Technologies, Methodologies, and Scholarship* (Cambridge: Cambridge University Press, 2008). A. Knowles has edited two volumes of essays covering a wide range of subjects in which historians have found GIS to be a useful tool: *Past Time, Past Place: GIS for History* (Redlands, Calif.: ESRI Press, 2002) and *Placing History: How Maps, Spatial Data, and GIS Are Changing Historical Scholarship* (Redlands, Calif.: ESRI Press, 2008). A more theoretical approach is offered by P. Ethington, "Placing the Past: 'Groundwork' for a Spatial Theory of History," *Rethinking History: The Journal of Theory and Practice* 11 (December 2007).

A number of studies applying GIS and spatial analysis to topics in the humanities have begun to appear over the past several years. An early work is William G. Thomas III and Edward L. Ayres, "The Differences Slavery Made: A Close Analysis of Two American Communities," *American Historical Review* 108 (2003): 1299–1308. Geoff Cunfer's prize-winning study of the Dust Bowl of the 1920s and 1930s is particularly noteworthy: *On the Great Plains: Agriculture and Environment* (College Station: Texas A&M University Press, 2005). Another good example is Brian Donahue, *The Great Meadow: Farmers and the Land in Colonial Concord* (New Haven, Conn.: Yale University Press, 2004). Although he does not discuss GIS, Brian Jarvis takes a spatial approach to cultural studies in *Postmodern Cartographies: The Geographical Imagination in Contemporary American Culture* (New York: Palgrave Macmillan, 1998). Finally, several Web sites offer interesting and useful applications of spatial analysis and spatial visualization: Valley of the Shadow: Two Communities in the American Civil War: http://valley.vcdh.virginia.edu/; Salem Witch Trials: Documentary Archive and Transcription Project: http://etext.virginia.edu/salem/witchcraft/; and Digital Roman Forum project: http://dlib.etc.ucla.edu/projects/Forum (all accessed 9 Jan. 2009).

GEOGRAPHIC INFORMATION SCIENCE AND SPATIAL ANALYSIS FOR THE HUMANITIES

In the field of geographic information science, unlike the humanities, the majority of the key references are published in academic journals rather than books. The name for the field was coined by Michael F. Goodchild in his oft-quoted 1992 essay published in the *International Journal of Geographical Information Systems* (*IJGIS*). *IJGIS* for many years was the premier journal in the field and it contains many seminal articles. Browsing the contents of this journal is always illuminating. It is interesting to note that *IJGIS* was renamed in 1996 to *International Journal of Geographical Information Science*, an illustration of the rapid incorporation of the new term into the academic language. Other key journals for GIScience articles include *Transactions in GIS* and *Cartography and Geographic Information Science* (renamed from *Cartography and Geographic Information Systems* in 1999). Many other journals across a wide range of disciplines regularly publish GIScience articles, including, for example, *Geographical Analysis*, *Computers & Geosciences; Communications of the ACM* (Association for Computing Machinery); *Environment and Planning B*; and *Computers, Environment and Urban Systems;* to name just a few.

A pair of two volume compendia by Maguire, Goodchild, & Rhind: *Geographical Information Systems: Principles and Applications* (London: Wiley, 1991) and (Longley, Goodchild, Maguire, & Rhind, 1999), affectionately known as "Big Book I" and "Big Book II," are the early pillars of most GIScience bookshelves. These contain foundational essays by many of the field's pioneers. A textbook by these editors, *Geographic Information Systems and Science* (London: Wiley, 2006) is also one of many well-stocked bookshelf standards. K. Kemp, *The Sage Encyclopedia of Geographic Information Science* (Thousand Oaks, Calif.: Sage Publications, 2008), gives a current snapshot of the key themes in the field with entries written by many of the field's academic leaders. Finally, a few books about GIS in the Humanities are beginning to appear. A very practical guide to using GIS is I. Gregory, *A Place in History: A Guide to Using GIS in Historical Research* (Oxford: Oxbow Books, 2003).

Spatial analysis as a comprehensive concept is much older than geographic information science. While examining the world from a spatial perspective has always been a core part of the discipline of geography, spatial analysis as an academic focus has its origin in the "quantitative revolution" in geography which took place during the late 1950s and 1960s. At that time much of the work in the field turned from description to quantitative approaches. This led to the development of many mathematical techniques which with the advent of computers has allowed spatial analysis to flourish. Early books on spatial analysis that continue to be useful classics include R. Chorley, *Spatial Analysis in Geomorphology* (New York: Methuen, 1972); P. Taylor, *Quantitative Methods in Geography: An Introduction to Spatial Analysis* (London: Waveland Press, 1977); D. Unwin, *Introductory Spatial Analysis* (London: Routledge, Kegan & Paul, 1981); and S. Fotheringham and D. Rogerson, *Spatial Analysis and GIS* (London: Taylor & Francis, 1994).

Like geographic information science, spatial analysis has been taken up by many disciplines so there are many books and hundreds of articles focusing on spatial analysis in various fields. Most GIS textbooks provide good overviews of spatial analysis. Recent comprehensive texts in spatial analysis include M. de Smith, M. Goodchild, and P. Longley, *Geospatial Analysis: The Comprehensive Independent Guide to Principles, Techniques and Software Tools* (Leicester: Troubador Books, 2008) (also available online at http://www.spatialanalysisonline.com/ga_book.html); S. Fotheringham and D. Rogerson, *Sage Handbook of Spatial Analysis* (Thousand Oaks, Calif.: Sage Publications, 2007); and D. O'Sullivan and D. Unwin, *Geographic Information Analysis* (Hoboken, N.J.: John Wiley & Sons, 2002).

EXPLOITING TIME AND SPACE

The theoretical nature of time and its relationship to space is a huge topic. Stephen Hawking's (relatively) readable *A Brief History of Time: From the Big Bang to Black Holes* (New York: Bantam, 1988) provides an overview from a physicist's perspective while the essays in Raymond Flood and Michael Lockwood, eds., *The Nature of Time* (London: Wiley-Blackwell, 1986) provide a variety of other approaches.

The importance of time in geography also has an extensive literature. Anne Buttimer's "Grasping the Dynamism of the Lifeworld," *Annals of the Association of American Geographers* 66 (1976): 277–92 and Doreen Massey's *For Space* (London: Sage Publications, 2005) and her paper that precedes it, "Space-Time, 'Science' and the Relationship between Physical Geography and Human Geography," *Transactions of the Institute of British Geographers: New Series* 24 (1999): 261–76, provide excellent overviews. John Langton's "Systems Approach to Change in Human Geography," *Progress in Geography* 4 (1972): 123–78 provides an alternative discussion from the perspective of systems' theory. The field of time geography is introduced in Nigel Thrift's *An Introduction to Time Geography* (London: Institute of British Geographers, 1977) or discussed in more detail in Don Parkes and Nigel Thrift's *Times, Spaces and Places: A Chronogeographic Perspective* (Chichester: John Wiley & Sons, 1980). The classic work in this field is Torsten Hagerstrand's "What about People in Regional Science?" *Papers of the Regional Science Association* 24 (1970): 7–21. Allan Pred's response "The Chronogeography of Existence: Comments on Hagerstand's Time-Geography and Its Usefulness," *Economic Geography* 53 (1977): 207–21 is a useful extension of this. Joe Weber and Mei-Po Kwan's "Bringing Time Back In: A Study on the Influence of Travel Time Variations and Facility Opening Hours on Individual Accessibility," *Professional Geographer* 54 (2002): 226–40 provides a more recent example of this kind of work.

Three useful books on how historical geography brings history and geography together are Alan R. H. Baker's *Geography and History: Bridging the Divide* (Cambridge: Cambridge University Press, 2003); Robin A. Butlin's *Historical Geography: Through the Gates of Space and Time* (London: Hodder Arnold, 1993); and Robert A. Dodgshon's *Society in Time and Space: A Geographical Perspective on Change* (Cambridge: Cambridge University Press, 1998).

There is an extensive literature on time in GIS. Gail Langran's *Time in Geographical Information Systems* (London: Taylor & Francis, 1992) was one of the earliest works, while Donna Peuquet's *Representations of Space and Time* (New York: The Guilford Press, 2002) provides a more up-to-date discussion. The essays in Max J. Egenhofer and Reginald G. Golledge, eds., *Spatial and Temporal Reasoning in Geographical Information Systems* (New York: Oxford University Press, 1998) provide alternative views.

A discussion on time in historical GIS is included in Ian N. Gregory and Paul S. Ell's *Historical GIS: Technologies, Methodologies and Scholarship* (Cambridge: Cambridge University Press, 2007). A number of papers and books in the field provide examples of how bringing space and time together using GIS allows new insights to be made in our understanding of history. These include Geoff Cunfer's *On the Great Plains: Agriculture and Environment* (College Station: Texas A&M University Press, 2005); Ian N. Gregory's "Different Places, Different Stories: Infant Mortality Decline in England & Wales, 1851–1911," *Annals of the Association of American Geographers* 98 (2008): 1–21; and Anne K. Knowles and Richard G. Healey's "Geography, Timing, and Technology: A GIS-Based Analysis of Pennsylvania's Iron Industry, 1825–1875," *Journal of Economic History* 66 (2006): 608–34.

QUALITATIVE GIS AND EMERGENT SEMANTICS

There are several ways to engage debates about genre difference between scientistic and humanistic discourses and the problem of semantics. One way is to follow the trail blazed by Mikhail Bahktin (*The Dialogic Imagination: Four Essays by M. M. Bakhtin*, trans. Caryl Emerson and Michael Holquist and ed. Michael Holquist [Austin: University of Texas Press, 1981] who, along with Zev Vygotsky (*Thought and Language*, rev. edition, ed. Alex Kozulin [Cambridge, Mass.: MIT Press, 1986]) pioneered theories about the manner in which language was inextricably bound up with contexts of use and how meaning-making through language is an ongoing transactional and reflexive process. Charles Bazerman explored the ways in which scientific language, and especially science writing, is a communicative practice in which meanings shift constantly as readers engage the written word from perspectives arising from their own situatedness (*What Written Knowledge Does: Three Examples of Academic Discourse Philosophy of the Social Sciences* 11: 361–88; *Shaping Written Knowledge: the Genre and Activity of the Experimental Article in Science* [Madison: University of Wisconsin Press, 1988]; *Languages of Edison's Light* [Cambridge, Mass.: MIT Press, 1999]). For Roy Harris "supercategories" of discourse, each constructed through its attachment to a certain area (e.g., art, science, history, religion), have their own semantics which are shaped moment to moment in the act of communication (*The Semantics of Science* [London and New York: Continuum, 1995]; *The Linguistics of History* [Edinburgh: University of Edinburgh Press, 2004]; *The Necessity of Artspeak: The Language of the Arts in the Western Tradition* [London and New York: Continuum, 2004]).

The ways in which these kinds of approaches to semantics have influenced the workings of the global Web have been surveyed with regard to both theoretical and technical issues (Karl Aberer, Philippe Cudré-Mauroux, and Aris M. Ouksel et al., "Emergent Semantics Principles and Issues," in *Database Systems for Advanced Applications Ninth International Conference* vol. 2973 of *Lecture Notes in Computer Science, Database Systems for Advanced Applications* (Berlin/Heidelberg: Springer, 2004), 25–38). Some scholars have envisioned the architecture of an emergent Web semantics based on linguistic theory (Sven Herschel, Ralf Heese, and Jens Bleiholder, "An Architecture for Emergent Semantics," in *Advances in Conceptual Modeling Theory and Practice* vol. 4231 of *Lecture Notes in Computer Science* [Berlin/Heidelberg: Springer 2006], 425–34).

In gauging GIS possibilities, there must be recognition that a key aspect in the construction of such architectures is the invention of a flexible ontology that can organizationally comprehend diverse metadata (N. Schuurman and A. Leszczysnki, "Ontology-Based Metadata," *Transactions in GIS* 11 [2006]: 709–26). Addressing that issue involves calculating how qualitative data and especially that which is shaped through social experience can comport with quantitative realism in GIS (N. Schuurman, "Reconciling social constructivism and realism in GIS" *ACME* 1 [2002] 75–90); S. Bell and M. Reed, "Adapting to the Machine: Integrating GIS into Qualitative Research," *Cartographica* 39 [2004]: 55–66). The role of visualization generally in the larger project of placing qualitative data within a georeferenced framework has been addressed in scholarship that offers examples of the possibilities for such visualization (L. Knigge and M. Cope, "Grounded Visualization: Integrating the Analysis of Qualitative and Quantitative Data through Grounded Theory and Visualization," *Environment and Planning* 38 [2006]: 2021–2037; M-P Kwan and J. Lee, "Geovisualization of Human Activity Patterns using 3D GIS: A Time-Geographic Approach," in *Spatially Integrated Social Science*, ed. M. Goodchild and D. Janelle [New York: Oxford University Press, 2004], 48–66). The prospect for multimedia as part of an enlarged effort at visualization was suggested by Rahim, Soomro T, Zheng Kougen, Turay Saidu, and Pan Yunhe ("Capabilities of Multimedia GIS," *Chinese Geographical Science* 9 [June 1999]: 159–65). Web projects that have experimented with multimedia in integrating qualitative historical data into a GIS framework include the *Salem Witch Trials Documentary Archive and Transcription Project*.

REPRESENTATIONS OF SPACE AND PLACE IN THE HUMANITIES

Understandings of space and place are intimately associated with landscape, a theme which unites many humanities disciplines and is also a good vehicle for exploring changing theoretical approaches as more widely applied. John Wylie's *Landscape* (London: Routledge, 2007) is a very good overview and readers who want a historical perspective can go back to the early writings of Carl Sauer, *Land and Life* (Berkeley: University of California Press, 1963) and W. G. Hoskins's *The Making of the English Landscape* (London: Hodder and Stoughton Ltd., 1954). The complexity of landscape is well illustrated by the collection of papers in Barbara Bender's *Landscape: Politics and Perspectives* (Oxford: Berg Publishers, 1993).

Over the last two decades discussions of landscape in many humanities subjects have often involved spatial technologies, especially Geographic Information Systems (GIS). A good example of the range of GIS applications, in this case within historical studies, is Anne Kelly Knowles, *Past Time, Past Place: GIS for History* (Redlands, Calif.: ESRI Press, 2002). The use of GIS has sometimes been contentious, for example in archaeology as illustrated by the papers in the collection edited by Gary Lock, *Beyond the Map: Archaeology and Spatial Technologies* (Amsterdam: IOS Press, 2000).

Theory plays a central role in many discussions of space, place, and landscape. For a good introduction to phenomenology, see Christopher Tilley, *The Materiality of Stone: Explorations in Landscape Phenomenology* (Oxford: Berg Publishers, 2004). For practice and dwelling the earlier works of Anthony Giddens, *The Constitution of Society* (Cambridge: Cambridge University Press, 1984) and Pierre Bourdieu, *Outline of a Theory of Practice* (Cambridge: Cambridge University Press, 1977) are still important starting points which Timothy Ingold builds on in *The Perception of the Environment: Essays on Livelihood, Dwelling and Skill* (London: Routledge, 2000). One response to the post-modernist crisis of representation is Nigel Thrift's "non-representational" theory in *Spatial Formations* (London: Sage Publications, 1996) brought up to date with the idea of "deep maps" in an intriguing book by Mike Pearson *'In Comes I': Performance, Memory and Landscape* (Exeter: University of Exeter Press, 2006).

Technology has developed alongside theoretical approaches, for example the claims of Virtual Reality (VR) in offering new understandings of space and place epitomized by the papers in Peter Fisher and David Unwin, *Virtual Reality in Geography* (London: Taylor & Francis, 2002), although see Mark Gillings' paper in that volume, "Virtual Archaeologies and the Hyper-Real: Or, What Does It Mean to Describe Something as Virtually-Real?" for a cautionary note. The Internet has added a whole new dimension to the theory and practice of space and place with new forms of social relationships being configured through virtual worlds and virtual communities. These changes inspired by Web 2.0 technologies could be profound, see Charles Leadbeater, *We Think: Mass Innovation, Not Mass Production* (London: Profile Book, 2008), and at http://www.wethinkthebook.net/book/home.aspx. Interesting spatial developments with great potential for the humanities include mashups with Google Earth leading the way, no books but join the blog at http://googlemapsmania.blogspot.com/#top.

MAPPING TEXT

One fundamental issue to GIScience is transformation of geospatial data to information. While some research questions remain, extensive GIScience research has resulted in Geographic Information Systems (GIS) tools and solutions to acquire data from georeferenced maps, remotely sensed images, sensors at environmental observation networks, and field surveys. For more information, readers may consult the essay by J. Jensen, et al., "Spatial Data Acquisition and Integration," in Robert B. McMaster and E. Lynn Usery, eds., *A Research Agenda for Geographic Information Science* (Boca Raton, Fla.: CRC Press, 2004).

Only until recently, GIScience researchers started exploring texts as an additional data source in GIScience. Nevertheless, GIScientists have a long history of studying natural languages to understand human spatial cognition and to improve GIS user interface design; see, for example, D. Mark and A. Frank, *Cognitive and Linguistic Aspects of Geographic Space* (Dordrecht, Netherlands: Elsevier, 1991); T. Nyerges, et al., eds., *Cognitive Aspects of Human-Computer Interaction for Geographic Information Systems* (Dordrecht, Netherlands: Elsevier, 1995); and M. Egenhofer, "Query processing in spatial-query-by-sketch," *Journal of Visual Languages and Com-*

puting 8:4 (1997), 403–24. Mark and Egenhofer showed nine common types of topological spatial relations (the 9-intersection model) between a line and a polygon; see "Modeling spatial relations between lines and regions: Combining formal mathematical models and human subjects testing," *Cartography and Geographical Information Systems*, 21:3 (1994), 95–212. The 9-intersection model has become the foundation for spatial information queries used in many commercially available GIS software packages.

Mapping text has at least three meanings relevant to GIScience: (1) *Geovisualization* that extracts locations in a text and presents a sequence of events over space, such as battle maps in the Valley of the Shadow project (http://valley.vcdh.virginia.edu/cwmaps1.html) and the Time-Map project (http://www.timemap.net/); (2) *Text2Sketch* coined by Max Egenhofer who suggested dependences between Natural-Language Spatial Predicates based on the 9-intersection model to enable the use of natural languages for geographic information retrieval; and (3) *Text analytics* that identifies important concepts in a text and structurally connects these concepts into readily searchable form or summative graphs for interpretation similar to computer automation of concept mapping or knowledge mapping; see R. Bose, "Advanced analytics: opportunities and challenges," *Industrial Management & Data Systems*, 109:2 (2009), 155–72; B. Gaines and M. Shaw, "Concept maps as hypermedia components," *International Journal of Human-Computer Studies*, 43:3 (1995), 323–61; A. O'Donnell, et al., "Knowledge Maps as Scaffolds for Cognitive Processing," *Educational Psychology Review*, 14:1 (2002), 71–86.

Beyond mapping, the use of text as a GIS data source not only demands a linguistic understanding of spatial relationships expressed in text as what is required in using natural languages in GIS user interface design, but it also challenges the ability to spatially reference "things" at locations. Such things can be concrete objects or geographic features, abstract geographic concepts (e.g., urban gentrification), or events and processes in geographic space; see M. Goodchild, "Geographical data modeling," *Computers & Geosciences*, 18 (1992), 401–408; M. Yuan, "Modeling semantical, temporal, and spatial information in geographic information systems," in M. Craglia and H. Couclelis, eds., *Geographic Information Research: Bridging the Atlantic* (London: Taylor & Francis, 1996), 334–47. The GIS database will thus provide the correspondent geographic coordinates to the place names or feature names in question.

Transformation of text to GIS data will enable a framework to integrate geovisualization, text2sketch, and text analytics, and thus will facilitate presentation, query, and analysis of texts in visual forms. This essay highlights three potential approaches to transform unstructured text to structured GIS database by spatialization, place-name matching, and geospatial inference. The development of robust GIScience methodology is necessary for a full scale GIS implementation of mapping the semantical, spatial, and temporal dimensions of rich texts in the humanities, so that we can visualize the multi-dimensionality of cultural and historical discourses. Furthermore, we will be able to develop data mining algorithms to elicit complex relationships embedded in text.

THE GEOSPATIAL SEMANTIC WEB, PARETO GIS, AND THE HUMANITIES

The digital online world of the Geospatial Web is changing rapidly and the printed word is invariably out of date before it can reach the reader. A search of the World Wide Web will generate countless references and applications for readers to explore. For a background history of mashups and the Geospatial Web, readers are recommended to see the work by Muki Haklay, Alex Singleton and Chris Parker, "Web Mapping 2.0: The Neogeography of the GeoWeb," *Geography Compass*, 2, 6, (2008), 2011–39. Andrew Turner's *Introduction to Neogeography* (Sebastapol, Calif.: O'Reilly Press, 2006) provides one of the earliest discussions of neogeography and makes a case for changing the emphasis from GIS and professional cartography toward a focus on making mapping tools that are available to non-expert users. The edited volume by Arno Scharl and Klaus Tochtermann, *The Geospatial Web: How Geobrowsers, Social Software, and the Web 2.0 are Shaping the Network Society* (London: Springer, 2007) also provides a wealth of information. The volume has extensive bibliographic references and some 25 chapters focused on developments in many aspects of the Geospatial Web, including foundations of the Geospatial Web, navigating the Geospatial Web, building the Geospatial Web, geospatial communities, and Geospatial Web services. Mike Goodchild's work, "Citizens as Sensors: The World of Volunteered Geography," *GeoJournal*, 69, 211–21, is illuminating for its perspective on the role of the amateur in geographic observation relative to more traditional spatial science. The papers given at a workshop on volunteered geographic information held under the auspices of the NCGIA in Santa Barbara in December 2007 are also helpful (http://www.ncgia.ucsb.edu/projects/vgi/).

GIS, E-SCIENCE, AND THE HUMANITIES GRID

Readers will realize from other contributors that GIS is relatively new to the humanities and, from this essay, will appreciate that e-science and grid computing are fresher still. Approaches combining these methodological approaches tend to exist online with almost no printed texts.

For a review of the potential of GIS and the importance of gazetteers, with some reference to grid computing, readers should refer to Ian N. Gregory and Paul S. Ell, *Historical GIS: Technologies, Methodologies and Scholarship* (Cambridge: Cambridge University Press, 2007). A short summary of the potential of e-science can be found in David Robey, "e-Science in the arts and humanities," *International Journal of Humanities and Arts Computing* 1:1 (2007): 1–3.

Richer sources are available online. The Arts and Humanities Research Council sponsored a number of briefing papers. Of particular relevance are Paul S. Ell, "Geographical Information System (GIS) e-Science: Developing a Roadmap" [n.d.] and other papers at Arts & Humanities e-Science Support Centre, *Briefing Papers:* http://ahessc.ac.uk/briefing-papers (accessed Jan. 2009). Also worth review is arts-humanities.net: Digital Arts and Humanities, *Beyond GIS: Geospatial Resources and Services for Scholars in the Humanities*: http://www.arts-humanities.net/briefing paper/beyond_gis_geospatial_resources_services_scholars (accessed Jan. 2009) and arts-humanities.net: Digital Arts and Humanities, *Grid:* http://www.arts-humanities.net/briefingpaper/grid (accessed Jan. 2009). These papers provide numerous links to other information.

Contributors

EDWARD L. AYERS, President, University of Richmond, and Professor of History, is best known for his prize-winning books on the American South, but he also is a pioneer in digital history through his development of The Valley of the Shadow Project at the University of Virginia, one of the web's most heavily trafficked history sites.

SUSAN BERGERON, Assistant Professor, Department of Politics and Geography, Coastal Carolina University, has research interests in geovisualization, historical geography and GIS, and utilizing advanced gaming graphics to generate virtual worlds. Her current work is focused on implementing an advanced serious game-based virtual world application capable of integrating humanities information within an immersive and interactive virtual landscape.

DAVID J. BODENHAMER, Executive Director, The Polis Center at IUPUI, and Professor of History, is author or editor of eight books and is co-author of a forthcoming book on humanities GIS. He has been involved with major humanities GIS projects in the U.S. and Europe, written numerous articles, and made presentations on the subject on four continents. He also serves as editor (with Paul S. Ell) of the *International Journal of Humanities and Arts Computing*.

JOHN CORRIGAN, Lucius Moody Bristol Distinguished Professor of Religion and Professor of History at Florida State University, is author or editor of ten books and co-editor of *Church History*. He has worked extensively in religion mapping and is a pioneer in the history of emotions, a field that engages concepts of space and place within the rubric of cultural studies.

PAUL S. ELL, Executive Director, Centre for Data Digitisation and Analysis, Queens University of Belfast, is author of a book on the spatial history of Victorian religion and, with Ian Gregory, a book on historical GIS. He also serves as editor (with David J. Bodenhamer) of the *International Journal of Humanities and Arts Computing*.

IAN GREGORY is Senior Lecturer in Digital Humanities at Lancaster University. His background is in using GIS in the humanities generally and in historical research in particular, subjects on which he has published widely. He has written two books: *Historical GIS: Technologies, Methodologies and Scholarship* (with Paul S. Ell) and *A Place in History*. He hosts the Historical GIS Research Network website (http://www.hgis.org.uk). He recently completed a pilot project on building a GIS of Lake District literature (http://www.lancs.ac.uk/mappingthelakes).

TREVOR M. HARRIS, Chair, Department of Geology and Geography, West Virginia University, and Eberly Professor of Geography, is one of the founders of the GIS and Society critique of spatial technologies and is an international figure in Participatory GIS, the semantic web and immersive GIS, a fusion of GIS with gaming and other advanced technologies.

KAREN K. KEMP, founding director of the GIS graduate program at the University of Redlands, is editor of the recently published *Encyclopedia of Geographic Information Science* and writes extensively on spatial analysis, with a special interest in the humanities.

GARY LOCK is Professor of Archaeology at the University of Oxford in the Institute of Archaeology and Department for Continuing Education and also a Fellow of Kellogg College, Oxford. He specializes in the use of computers in archaeology, especially GIS. In addition to many papers and book chapters, he has published *Using Computers in Archaeology: Towards Virtual Pasts; Archaeology and Geographic Information Systems: A European Perspective;* and *Beyond the Map: Archaeology and Spatial Technologies*, among others. He has also carried out major excavations on later prehistoric and Roman sites in Oxfordshire. His current long-term excavations are at Marcham/Frilford.

L. JESSE ROUSE, Assistant Professor of Geography at the University of North Carolina at Pembroke, focuses on the intersection of Geographic Information Science and phenomenology to address social science and humanities research questions.

MAY YUAN, Professor of Geography, University of Oklahoma, is the North American editor of the *International Journal of Geographic Information Science* and co-author of *Visualization and Computation of Dynamics in Geographic Domains*. Her research centers on geographic dynamics and spatiotemporal representation of physical, social, and recently cultural processes from a wide range of data sources, including texts.

Index

Page numbers in italics refer to figures.

Access grid. *See* Grid technologies
Accuracy, 8, 31, 43, 96, 104, 152, 156, 169; quantitative, 77
Aggregation, data, 68–69
Alexandria Digital Library, 114–115
AlphaWorld, 100
Ambiguity, ix, xiii, 23, 28, 76–78, 83, 87–88, 126, 169
Analysis, 1–2, 4, 6, 8, 18, 28, 47, 54, 56, 67, 69, 72, 76, 77, 79, 85, 87–88, 92, 95–96, 101, 113, 117–118, 121, 135, 147, 150; clustering, 111; DEM-based, 99; diachronic, 65, 66; document, 120; geographical, 1; GIS, 111; layer-based, 94; network, 94; of patterns, 92; semantic, 110, 116, 119; spatial data, 53; spatio-temporal, 73; statistical, 69; synchronic, 65; territorial, 90; text, 119, 121, 188; topological, 101; time-series, 72–73, 144
Analysis, raster (cell-based), 53, 95; cost-surface, 95; least-cost paths, 95; line-of-sight, 95; map algebra, 52, 53; trend surface, 54; viewsheds, 53, 95–96
Analysis, spatial, x, 17, 31, 32, 47, 48–56, 55, 73, 111, 120, 126, 128, 169, 173, 179–181; buffer, 50; overlay, 50, 51, 52, 53; point pattern, 53; raster (cell-based), 53, 95; shortest path, 53; site suitability, 50–52, *52*; statistical, 54; vector, 53, 94; visual, 48–49, 111
Animations, 24, 72–73, 102, 120

Annales school, 20
Annalistes, 16, 21
Application Programming Interface (API), 103, 127, 132
ArcGIS, 160
ArcGIS Explorer, 139, *139*
Archaeologists, vii, 21, 24–25
Archaeology, 21, 64, 90, 92, 100, 104–105, 124, 185
ArcInfo, 17
ArcView, 81
Areal interpolation. *See* Interpolation, areal
Artifact(s), viii, 21, 26, 84–85, 169; cultural, 26–28
Artificial intelligence, 85, 87
Arts and Humanities Research Council (U.K.), 146, 163, 189
Asynchronous Javascript And XML (AJAX), 132
Atlantic World, 25, 85
Atlases, 91–92, 117, 124, 140
Attribute(s), 20, 40–41, 45, 48, 51–52, 55, 59, 72, 78, 82, 85, 93–95, 100–101, 113, 121, 126, 140, 145, 169; tables, 41, *41*, 45, 55; value categories, 41–42

Bakhtin, Mikhail, 4
Bioinformatics, 110–111
Bourdieu, Pierre, 5, 185
Buffer, 50, *51*, 53, 68, 93
Buffering, 93

Canadian GIS, 17
Cartograms, 94
Cartography, 43, 90, 99, 130, 188; analytical, 90; automated, 17; digital, 130; representational, 90
Causation, xiii, 8, 10, 174
CAVE (Cave Automatic Virtual Environment), 25, 97
Census(es), x, 21, 49–50, 54, 66, 113, 114, 143, 149, 152–154, 163; historical, 124; Irish, 153–154. *See also* Data, census
Certeau, Michel de, 15
Chavin, Peru, 103–104
Chorley Report, 89
Chronotope, 4
Cityscape, viii, 21
Common Ground, 92; Parish Mapping Project, 92, 104
Communities, virtual, 104, 132, 186
Complexity, viii–ix, xiv, 18, 23–24, 28, 32, 35–36, 58, 66, 80–81, 96, 105, 124, 168–169, 185
Computational grid. *See* Grid technologies
Context, 1, 3, 12, 15, 20, 29, 130, 135, 140, 183; geographical, 28, 109, 119; historical, 8, 54; spatial, 17, 25, 45, 103
Context and Relationships NEH project, 155, 158, 164
Contextualization, 110, 155; of place names, 116; of space, 130
Contingency, ix, 26, 28; deep, 6–7, 12, 174–175; surface, 6
Coordinate(s), systems, 169; Cartesian, 20; Latitude/Longitude, 43; National Grids, 43; State Plane, 43; Universal Transverse Mercator, 43
Cronon, William, 64–66
Cunfer, Geoff, 21, 72, 179, 183

Data: attribute, 72, 82, 93–95, 100–101; census, 36, 37, 67, 94, 143–144; clearinghouse, 159–161; complex, 80; consumers of, 127–129, 136, 140; dynamic, 80; geocoded, 85–87; geographic, 32, 54, 121; geospatial, 117, 127, 135, 159–160, 186; GIS, 55, 111, 118–119, 121, 134, 137, 188; historical, 136, 184; humanities, 76, 82, 129, 134, 145; models, 46–47, 50, 59, 73; multimedia, *see* Multimedia GIS (MMGIS); non-geographic, 111; open, 134, 137; producers of, 127–129, 136, 140; qualitative, 80, 85, 87, 171, 184; quantitative, 18–21, 29, 59, 69–71, 144, 148, 176; querying, 77; raster, 47, 48, 51–52, 59, 95; sharing, 55, 134, 146, 164, 173; sources, 127, 134–135, 186–187; spatial, xi, 2, 18, 50, 53, 56, 66, 88–89, 101, 104, 126–127, 129, 131, 135, 140, 167, 171; statistical, 144, 159, 161, 163; tagging of, 85–86; temporal, 66; topological, 80; user-generated, 135; vector, 47, 47, 50–51, 83, 93, 132, 138
Data grid. *See* Grid technologies
Data mining, 146, 172, 188; spatial, 53
Database(s), 8, 20, 23, 37–38, 40, 42, 46–47, 61, 67, 69–70, 94, 103, 105, 124, 135, 147–148; design of, 40; digital, 172; GIS, 40, 109, 111, 119–120, 187–188; humanities, 134; schema, 40, 169, 174
Datasets, 17, 52; GIS, 45, 72, 77, 81, 84–85, 121, 132, 134, 137, 149, 160, 164; humanities, 147, 159; online, 147; user-generated, 132
Datum, 43, 79, 169; geodetic, 43; North American Datum 1983 (NAD83), 43; World Geodetic System 1984 (WGS 84), 43
Deep contingency. *See* Contingency, deep
Deep mapping. *See* Mapping, deep
Deep maps, 27–28, 90, 100, 174
Density, 9, 38; dot, 72; population, 36, 37, 52, 68; semantic, 112, *112*; spatial, 72; temporal, 72
Density smoothing, 72
Determinism: environmental, viii, 90; technological, 95
Digital Elevation Model (DEM), 95–96
Digital history. *See* History, digital
Digital humanities, 58, 129, 134

Digital Roman Forum Project, 24–25
Digital Terrain Model (DTM), 95

E-content, 164
Einstein, Albert, 58, 61–62
Electronic Cultural Atlas Initiative (ECAI), 81, 102, 117, 158–161, *158*, 164–165
Emergent semantics, 76, 81, 84–86, 88, 176, 183–184
Emotion, 175; history of, 83
Empiricism, 18, 29, 90–91
Endurants, 93–94
English Place-Names Society (EPNS), 150–151, *151*, 157
Entities, 18, 34–35, 38–39, *38*, 45, 48, 61, 113–114, 169
Environmental determinism. *See* Determinism, environmental
Environments: immersive, 25; virtual, 24–25, 176
Epistemology, 18–19, 24, 29, 109, 176; geographical, 109; GIS and, 18–19, 24, 170; positivist, 19, 170
E-research, 145
E-resources, 143, 149–150, 154, 160–161, 163–164, 172; humanities, 147–149, 160; online, 147; statistical, 144; text-based, 144
Error, 55–56, 71, 152; and uncertainty, 55–56
E-science, 143–146, 148, 150, 152, 160, 164, 172, 173, 189; GIS, 162, 165; humanities, 161, 148, 153–154, 158–161, 163
Euclidean: coordinate system, 169, 172; geometry, 19, 59; space, 66
Event, 4, 6, 9–10, 32; and process, 4, 6–9
Experience, viii, 1, 109, 119, 169, 172, 175; American, 14; human, xii, 3–4, 8, 24–25, 62, 65, 72, 84, 90, 98, 100, 109–110, 117, 172; shared, 104; social, 184; texture of, 1, 169; user, 23, 97, 131
Exploratory Data Analysis (EDA), 73
Exploratory Spatial Data Analysis (ESDA), xi, 53

eXtensible Markup Language (XML), 115, 132–133

Fields, data, 35, 38, 41, 46–48, 169; pseudo, 36
Folksonomy, 104
Foucault, Michel, 15
Fuzziness, 19
Fuzzy set theory, 36

Gaddis, John Lewis, xi
Gazetteer(s), 27, 45, 70–71, 114, 119, 121, 124, 135, 140, 152, 155, 158, 162, 164, 189; administrative, 150; census-based, 153; digital, 114–116, 119, 121, 172; event, 118–119, 121; historical, 149, 161; online, 87; place-name, 149–150, 154, 157, 161
Gazetteer-matching, 111
Geertz, Clifford, 15. *See also* Thick description
Generalization, xii, 3, 11, 77, 169, 174
Geobrowsers, 115, 127, 159–160
Geocoding, 76, 134
Geographic information. *See* Information, geographic
Geographic Information Science (GIScience or GISci), ix, 31–33, 35, 38, 40, 125, 134, 170, 174, 178, 180, 186–188
Geographic Information Systems (GIS), vii–viii, 2, 9, 16, 21, 25, 27–28, 31–32, 34–37, 42–43, 45–48, 53–56, 58–59, 69–73, 76–79, 86, 89, 91–93, 110, 124, 126, 139, 145, 159, 167, 174, 178, 181, 185; archaeology and, 90–91; contextual, 134; Critical, ix, 19–20, 128, 169, 177–178; and e-science, 150, 162–165; functionality, 94–95, 101, 130, 148; historical, *see* Historical GIS; history of, 17–20, 178; humanities, *see* Humanities GIS; immersive, xiv; issues of time in, 23, 66, 85, 173, 182; memory and, 26–28; multimedia, *see* Multimedia GIS (MMGIS); Pareto, *see* Pareto GIS; Participatory (PGIS), 128–129,

131, 140, 169; phenomenology and, 175; political economy of, 128, 130; positivism and, ix–x, 18–19, 81, 168, 170; qualitative, 76, 184; quantitative, 77, 130, 169; and Society, ix, 20, 125, 127–128, 169, 188–189; spatio-temporal, xi, 70, 173; structural knowledge distortion and, 128, 170; virtual, 171; Web, 100–101
Geographic Markup Language (GML), 115–116
Geography, discipline of, ix, 2–3, 5–7, 14; administrative, viii, 149; census, 152–154; critical, xv, 1–2, 5; cyber, xi, 24, 171; historical, 22, 63, 65, 124, 144, 182; and history, *see* History, and geography; and humanities, xv; natural, 15; of place, 2; shared, vii, ix; of space, x; spatial turn and, 1; time, 63
Géohistoire, 20, 22
Geo-inference. *See* Inference, geospatial
Geoparsing, 115–116, 135, 140, 172
Georeferencing, 70–71, 109–110, 117, 172
Geospatial inference. *See* Inference, geospatial
Geospatial markers. *See* Markers, geospatial
Geospatial Semantic Web, 120, 124, 127, 129–130, 134, 140–141, 172
Geospatial Web, 102, 127, 130–131, 133, 135–136, 139–141, 171–172, 188
Geostatistics, 54; kriging, 54
GeoTime, 120
Geovisualization, xi, 24, 101, 167, 171, 176, 187–188
GeoXwalk, 115
Giddens, Anthony, 5, 15, 98, 185
GIS knowledge base, 116–118
GIS 2.0, 128
Global Positioning Systems (GPS), viii, 21, 43, 55
Google: Scholar, 157; SketchUp, 132; 3D Warehouse, 132–133
Google Earth, 34, 56, 103, 116, 127, 131–132, 139, 186; Chavin example, 103–104; Global Awareness layer, 131

Google Maps, 56, 103, 127, 131–132, 139; Application Programming Interface (API), 136
Great Britain Historical GIS. *See* National Historical GIS, Great Britain
Grid computing, 189
Grid technologies, 145–146, 173; access grid, 146; computational grid, 146–147; data grid, 146–149, 151,161, 164, 173

Hagerstand, Torsten, 63
Harvard Laboratory for Computer Graphics, 17
Hermeneutics, 2, 77
Historic Landscape Characterisation (HLC), 93
Historical GIS, ix, 21–22, 44, 66–67, 81–82, 124–125, 153, 161, 163, 173, 179, 185; Dust Bowl case study, 72; infant mortality case study, 67–69; space and time in, 66–69, 182
History, 3–7, 28, 62, 91, 101, 104, 109, 124–125, 183; and archaeology, 21; communal, 26; digital, 22; and geography, 3, 7–8, 16, 20, 58–59, 63–64, 173, 182; GIS and, *see* Historical GIS; intentional, 25; interpretive, 9, 174; landscape and, 91; and memory, 24, 29; methods, 70; and place, 3, 29; postmodernism and, 29; quantitative, 70, 72; scientific, 29; spatial, 22; and spatial turn, 3
History Engine, 8–9
Hoskins, W. G., 91, 185
Humanist turn, 105
Humanities: digital, *see* Digital humanities; GIS and, *see* Humanities GIS; postmodern, xv; scholarship, xiii–xiv, 16, 23, 58, 126, 134, 167, 170, 172; spatial, *see* Spatial humanities
Humanities computing, 129
Humanities GIS, viii, xiv, 27–29, 40, 73, 77, 82, 85, 88, 93, 96, 99, 109, 119, 125, 127, 129–130, 133, 136, 141, 144–145, 148, 161, 164, 168, 171, 176, 179–180

Immersion, 97
Inference, geospatial, 117–121, 188
Information: geographic, vii, 31–32, 42, 117, 130, 187; qualitative, 31; quantitative, 144; spatial, 11, 130, 140, 170, 187
Infrastructure, 148; e-research, 145; information, 164; place-name, 150; spatial data, 21
Inhabitation, 91
Interdependency, xii
Interoperability, 115, 133, 135, 139, 163–164, 173
Interpolation: areal, 49–50, 49, 52, 68; spatial, 112
Interpretation, 72

Keyhole Markup Language (KML), 116
Knowledge, local 136; production of, 18; theory of, 18
Knowledge base, GIS, 116–118, *118*
Knowledge discovery, vii
Kuhn, Werner, 93

Landscape(s), x, xi, 16, 21, 35, 38, 48, 54, 89–99, 104–105, 133, 136, 150, 175, 185; cultural, 6, 26, 28, 33, 91, 99, 124; of culture, 28; data, 21; historical, 33, 136, 138; inhabited, 95; as metaphor, 105; past, 21; physical, 26, 91; and place, 14, 28; social, 33; spatial, 3; temporal, 3; virtual reconstructions of, 132, 139
Landscape architecture, 17
Langran, Gail, 66
Langton, John, 65–66
Language, 85; spatial, 109
Latent Semantic Analysis, 119
LiDAR (Light Detection and Ranging), 96
Linguistics, 70–71, 110, 155; computational, 156
Local knowledge. *See* Knowledge, local
Location, absolute, 42; direct, 43; indirect, 43; measured, 126; relative, 42; spatial, 31

Location-based Services, viii, 17, 131
Logic, Boolean, 19

MacCarthy, Cormac, 15
Map algebra, 52; functions, 53–54
Mapping: cartographic, 169; community, 128; conceptual, 169; counter, 128; crime, 72; deep, xiv, 26–27, 174–176; geographic, viii, 110, 116; GIS, 110, 126, 128, 140, 171, 175; humanistic, 175; interactive, 102, 135; Internet, ix, 135; thematic, 90; value-laden, 175; Web, 127, 131–132, 135–138
Maps: cartographic, 169; choropleth, 72, 94; cinematic, 8; deep, 27–28, 90, 100, 174, 185; dot-density, 72; dynamic, 11–12; geographical, 22, 112; georeferenced, 186; historic, 101; and history, 3; as metaphors, 109; motion, 10; self-organizing, *see* Self-organizing maps (SOM); static, 12; topographic, 47–48, 92
Markers: chronological, 121; geospatial, 111, 118–119, 121, 172
Mashup(s), 100, 102–103, 127, 132, 135, 186, 188
Massey, Doreen, 65–66, 72, 173, 178, 181
McHarg, Ian, 50
Memory, viii, 20, 24–25, 28, 169, 174; communal, 26; dynamic, 28; personal, 26; and place, 25, 27, 29; public, 26; social, 26; structuring of, 26
MESH WebGIS, 101
MetaCarta, 140
Metadata, 147; ECAI, 159, 160
Microsoft Virtual Earth, 56, 127, 131
Minkowski, Hermann, 61–62; light cones, 62
Modeling: agent-based, 97; of movement, 95; predictive, 90; spatial, 54; of visibility, 95
Modifiable Areal Unit Problem (MAUP), 36, 50
Motion map. *See* Maps, motion
Multimedia, 9, 28, 82–83, 87–88, 131, 134–139, 159, 161, 176, 184; audio as, 82, 85, 100, 131–132, 136, 138–139;

embedded,132, 135; images as, 9, 85, 87, 100–101, 103, 128; local knowledge as, 128, 136; narrative as, 128; photographs as, 25, 82, 103–104, 132, 134, 136, 138, 161; sketch maps as, 128; spatial, xi, 24, 76, 83, 130, 176; text as, 12, 82, 85, 128; video as, 12, 82, 87, 100, 103–104, 131–132, 138–139, 161
Multimedia GIS (MMGIS), 76, 80–84, 87, 128, 130, 184
Multiplicity, ix, 28
Multi-scalarity, 92

Narrative, xiii, 1, 3–4, 7, 9, 12, 26, 29, 121, 126, 128, 132, 169, 172, 174, 176; counter, 15; expert, 25, 28; in GIS, 109–111, 116
National Geospatial Digital Archive (NGDA), 115
National Historical GIS, 21, 67, 153; China, 21; Germany, 21; Great Britain, 21, 67, 81, 161; Russia, 21; United States, 21
Natural language, 11, 186–187
Natural language processing, 110, 115, 172
Neogeography, 127, 140, 171, 188
New Historicism, ix, 1
New media, 131, 135
Nuance, xiii, 23

Objectivity, 29, 90–91
Objects, 10, 18–19, 25, 35–36, 38, 43, 50, 100, 132, 159, 169, 187; discrete, 35; geo-, 94; represented in GIS, 46–48; spatial, 18, 95; 3-dimensional, 64
Observation, 19, 85, 90–91, 94, 170, 175
Ontologies, 38, 40, 46, 48, 135
Ontology, 18, 38–39; in computing, 38–40, 39, 84; GIS, 76, 88, 184; geographical, 109
Ortner, Sherry, 5, 7

Pareto GIS, 127, 130, 136; Morgantown, W.Va., case study, 136–138
Pareto principle (80:20 rule), 129–130
Patterns: spatial, vii, 73, 121; temporal, 73

Perception, 96
Perdurants, 94
Phenomenology, 90–91, 98, 175, 185; 'being-in-the-world,' 98; dwelling, 98, 185; Heidegger and, 98; Merleau-Ponty and, 98
Pickles, John, 18, 178
Place(s), viii, xi, 1, 5, 20, 26, 58, 63, 89, 100, 126, 132, 173–175; deep mapping of, 176; geography of, 2; GIS and, 28, 59; History and, 3; inhabited, 91; landscape and, 14, 16; and memory, *see* Memory, and place; notions of, 5; past, 21, 63; qualitative aspects of, 126; quantitative aspects of, 126; representations of, *see* Representation(s), of place; sense of, 6, 24, 91, 99, 136, 139, 175; and space, *see* Space(s), and place; spirit of, 27; and time, *see* Time, and place; understanding, 92, 97, 121, 140
Place-making, 1, 173–174
Place names, 45, 70–71, 104, 114–116, 118–119, 121, 135, 149–158, 163, 172, 187; contextualizing, 116; gazetteers, *see* Gazetteer(s), place-name; geocoding, 134; matching, 115–116
Platial, 132
Postmodernism, 16, 29, 90, 103
Practice theory. *See* Theory, practice
Precision, xiii, 20, 28, 31, 43, 48, 83, 104, 169
Process, and event, 4, 6–9
Projection, 44–45, *44*, 113, 169; map, *44*
Proximity, 23–24, 94, 105, 112; geographic, 114; spatial, 119

Qualitative GIS. *See* Geographic Information Systems (GIS), qualitative
Quantitative GIS. *See* Geographic Information Systems (GIS), quantitative
Quantitative revolution, 90; in geography, 181

Realism, 19; quantitative, 184
Reconstruction, virtual, 25, 132–133, 137, 139, *139*

Reductionism, xiii, 90, 169
Relationships, 18, 38, 40, 59, 167, 170; causal, 10, 48; complex, 188; data, 81, 86; dynamic, 176; geometric, 47; power, 91; qualitative, 91–92; semantic, 112, 133–135; social, 94, 100, 186; spatial, vii, 4, 11, 92, 98, 110, 116, 119–121, 126, 171–172, 187; temporal, 4, 11; topological, 11, 117
Representation(s), 3, 7, 32–33, 55–56, 66–67, 83, 90, 96, 99, 115, 128, 132, 185; of behavior, 97; Cartesian, 91; cartographic, 130, 171; computer, 39, 47; of culture, 87, 176; data, x, 128; digital, 46, 95, 105; dynamic, xi, 29; of features, 126; geographic, 97; GIS, 22; of inhabitation, 98; local, 16; of memory, 29; multidimensional, 24; official, 19, 103; of the past, xi; of phenomena, xi; of place, xi, 29, 89, 104, 115; place-based, 97; raster, 47–48, 95–96; and scale, 90, 92–95; social, 100, 102, 105; of space, x, 5, 14, 59, 61, 89, 94, 104, 170; space-based, 91; spatial, 89, 102, 170; symbolic, 89; temporal, 1, 93; of time, 61; vector, 47–48; virtual, 171; of visibility, 96; visual, 134
Representative fraction, 33, 34
Resource Description Framework (RDF), 133

Sahlins, Marshall, 5
Said, Edward, 15
Salem Witch Trials project, 22, 81, 87, 184
SAS TextMiner, 119
Sauer, Carl, 91, 124, 185
Scale, xi, xiii, 23, 33–34, 34, 46, 48, 78, 85, 90, 92–93, 95–96, 99; analytical, 92; conceptual, 92; phenomenological, 92; representational, *see* Representation(s), and scale
Science: environmental, viii, 17, 27, 54; geographic information, *see* Geographic Information Science (GIScience or GISci); information, 31, 76; spatial information, 125, 170

Second Life, 25, 100, 105
Selectivity, xiii
Self-organizing maps (SOM), 111–112, 114, 172
Semantic Web, 129, 133–134, 154–155. *See also* Geospatial Semantic Web
Semantics, 37, 40, 84, 104, 119, 135, 183–184; emergent, 84–86, 88, 176, 183–184; humanities, 84; reocentric, 84
Sense of place. *See* Place(s), sense of
Serious gaming. *See* Technology, gaming
Service Oriented Architecture (SOA), 129, 134
Simultaneity, xiii, 10, 28
Situatedness, 84, 183
SketchUp. *See* Google, SketchUp
Snow, C. P., viii, 109–110
Snow, John, vii, 76; London cholera map, vii, 76
Space(s), x, xiv, 1, 15–16, 58, 65, 85, 124; absolute, 172; attribute, 112–113; body and, vii, 21; Cartesian, 20, 91; characteristics, 114; cognitive, 9; collaboratory, 131–132; concepts of, vii, 63, 66, 170, 179; conceptual, 29; contextualizing, 130; coordinate, 169; data, 112, *112*; definition of, 20, 24; emotional, 21; gendered, vii; generalized, 5, 174; geographic, vii, 14, 20–22, 24, 32, 43, 112–114, 121, 126, 134, 167, 172, 187; geography of, x; geometric, 19; GIS and, 20, 22, 24, 58–59, 126, 169, 173–174; imagined, 14, 20; meaning of, 16, 20; metaphorical, 172; notions of, 14; objective, 176; physical, 21; and place, viii, 1, 6–7, 9, 16, 24, 78, 89, 104–105, 128, 141, 174, 176–179, 185–186; production of, 16, 170; public, 26; racialized, vii; real, 14, 29; relational, 20; relative, x, 172; religion and, 77–78; representations of, *see* Representation(s), of space; semantic, 112, 121, 172; theories of, 3; and time, *see* Time, and space; 3-D, 43; understanding of, 2, 104
Space-time, 58, 60–62, 64–65, 72, 173

Spatial analysis. *See* Analysis, spatial
Spatial autocorrelation, 46, 54, 114
Spatial data analysis. *See* Analysis, spatial data
Spatial data mining. *See* Data mining, spatial
Spatial history. *See* History, spatial
Spatial humanities, vii–ix, xiv, 36, 45, 54, 93, 167–176, 178
Spatial interpolation. *See* Interpolation, spatial
Spatial modeling. *See* Modeling, spatial
Spatial multimedia. *See* Multimedia, spatial
Spatial primitives, x, 91, 126, 128, 139, 169
Spatial query, 24
Spatial storytelling, 126, 140, 172
Spatial turn, vii, 173; in humanities, viii; in history, 3
Spatialization, 53, 135; of text, 111–114, *113*, 120, 172, 188
Subjectivity, 29, 87, 94, 176
Systems theory. *See* Theory, systems

Tagging, 104; collaborative, 104; geospatial, 120; geotagging, 130; linguistic, 85; retagging, 85–86; social, 104
Taskscape, 91, 99, 175
Taxonomy, social. *See* Folksonomy
Technology, gaming, 25, 28, 100, 171; geospatial, 125; spatial, x, 21, 185
Temporal turn, viii, 1, 12, 173
TerMine, 119
Text(s), 29, 45, 70, 83, 101, 103, 110–111, 134–135, 144, 147, 150, 154, 169–170, 172, 186; analysis, 119, 121; converting to GIS data, 70, 111, 118, 120–121, 187–188; encoding, 154–156, 163; geoparsing, *see* Geoparsing; georeferencing, 71, 73; mapping, *see* Text mapping; mining, *see* Text mining; as multimedia, 82–83, 85, 88, 128, 132, 136–137, 139, 159; spatialization of, *see* Spatialization, of text; tagging, 104, 120; unstructured, 111, 113, 172, 188; visual analysis of, 121, 188

Text analytics, 110, 188
Text Encoding Initiative (TEI), 154–156, 163
Text mapping, 71, 109–110, 113, 120, 187
Text mining, 110–111, 120, 154, 172–173; tools for, 111, 119, 121
TextAnalyst, 119
Text-map transformation, 109–114, *113*, 118–119, 172, 188
Text2sketch, 187–188
Theory: agency, 98, 175; hierarchy, 113; linguistic, 184; more-than-representational, 99; non-representational, 98, 99; practice, 4–8, 91, 98–99, 175, 178, 185; structuration, 98; systems, 65, 182
Thick description, xi, 100, 174–175
'Third culture,' 109–110
Thrift, Nigel, 98, 182, 185
Thunen, Johann von, 64–65
Time, x, 1, 3–5, 15, 56, 58, 60, 66, 85, 90, 93, 178, 181; branching, 60; calendar, 60; conceptualizing, 60, 94; container, 60; cyclical, 60; as dimension, 61; discrete, 59; event, 61; GIS and, 23, 59, 66, 71, 81–82, 128, 173, 182; and history, 6, 16, 62–64; and humanities, 64, 173; linear, 60; maps of, 10; and motion, 10, 82; multiple perspectives of, 60; Newtonian model of, 61; passage of, 10; and place, xi, 3–4, 7–9, 174; practice theory and, 5; real, 61; representations of, 61; and scale, 93; sense of, 6; and space, x–xi, 3–4, 7–10, 12, 15, 22, 27–28, 58–59, 61–67, 70–73, 85, 102, 131, 173–174, 181; space-time, *see* Space-time; theme and, 58, 67, 71; valid, 61
TimeMap, 81, 102, 120, 159–161, 187
Timescapes, 3
Time-series analysis. *See* Analysis, time-series
Tobler, Walter, First Law of Geography. *See* Spatial autocorrelation
Topology, x, 16, 47, 59, 126, 169
Travel accounts, 27, 64, 144, 161
Turner, Frederick Jackson, 15, 64–65

Uncertainty, x, xiii, 19, 23, 28, 55–56, 103–104, 169; and error, 55–56
Urban planning, 17
User generated content, 130–132, 135, 138

Valley of the Shadow project, 22, 179, 187
Viewshed analysis. *See* Analysis, raster (cell-based)
Virtual Earth. *See* Microsoft Virtual Earth
Virtual GIS. *See* Geographic Information Systems (GIS), virtual
Virtual globes, 127, 131–132, 135, 139
Virtual Jamestown, 25
Virtual reality, xi, 24–25, 82, 97, 100, 171, 186
Virtual worlds. *See* Worlds, virtual
Vision of Britain Through Time, 149, 153, 158, 161–162, 162, 164–165
Visualization, 17, 21, 24, 72, 76, 82–83, 86–87, 99, 102, 144–146, 179, 184, 187–188; cinematic, 9; data, 94; geographic, 97; information, 111, 165; pseudo-3-D, 95, 132; of scale, 85; 3-D, 21, 24, 82, 136, 140; time-series, 144; 2.5-D, 97, 171. *See also* Geovisualization
Volunteered Geographic Information, 127, 129, 141, 189
VRGIS, 97

Web mapping, 132, 135–136, 188; applications, 127, 131–132, 137–138; technologies, 131
Web services, xi, 127, 129, 132; geospatial, 188; GIS-enabled, 24, 116
Web 2.0, xv, 102–104, 127, 129–130, 141, 150, 156–157, 164, 186; geobrowsers and, 127; mashups and, 135; and Semantic Web, 154–155
We-think culture, 102, 104
Williams, Raymond, 5, 7
World Heritage Sites, UNESCO, 103
Worlds: material, 19, 92–94, 98–99, 175; social, 92–93; virtual, 21, 24–25, 28, 90, 100, 105, 186

YALE text mining software, 119

Z39.50 client-server protocol, 148, 164

Printed and bound by CPI Group (UK) Ltd, Croydon, CR0 4YY

09/06/2025

14685932-0001